Out of the Woods

ALSO BY LYNN DARLING

Necessary Sins

A Memoir of
Wayfinding

Out of
the
Woods

Lynn
Darling

HARPER

www.harpercollins.com

HarperCollins books may be purchased for educational, business, or sales promotional use. For information, please e-mail the Special Markets Department at SPsales@harpercollins.com.

Grateful acknowledgment is made for permission to reproduce from the following:

Excerpt from "The God Abandons Antony" from *Collected Poems* by C. P. Cavafy. Translated by Edmund Keeley and Philip Sherrard. Copyright © 1992. Reprinted by permission of Princeton University Press.

Excerpt from "The Seven Sorrows" from *Collected Poems* by Ted Hughes. Copyright © 2003 by the Estate of Ted Hughes. Reprinted by permission of Farrar, Straus and Giroux, LLC.

Excerpt from "Winter in the Village" from *Collected Poems* by Ted Hughes. Copyright © 2003 by the Estate of Ted Hughes. Reprinted by permission of Faber and Faber, Ltd.

FIRST EDITION

Library of Congress Cataloging-in-Publication Data has been applied for.

ISBN 978-0-06-171024-7

14 15 16 17 18   OV/RRD   10 9 8 7 6 5 4 3 2 1

For Dorothy Elizabeth Budnik Darling

Often and often it came back again
To mind, the day I passed the horizon ridge
To a new country, the path I had to find
—EDWARD THOMAS, "Over the Hills"

Perhaps, being lost, one should get loster.
—SAUL BELLOW, *Humboldt's Gift*

Out of
the
Woods

Prologue

Getting lost is easily avoided, say people who never get lost. Pay attention along the way. Keep track of the sun and the shape of the horizon. Turn around, every now and then, and look back at where you've been. Remember landmarks. Keep in mind your panic azimuth. Take your compass, take your bearings, take your time.

But sometimes, you don't pay attention. Sometimes, beguiled by the beauty of the passing moment, you walk along the path you chose a long time ago without noticing the subtle turn it has taken, the darkening sky, or the slight rustle in the leaves that means the wind is coming up. Something disturbs your mazy thoughts: a movement, swift and silent, catches at the corner of your eye. A twig snaps. The shadows begin to steal uncomfortably close. You quicken your pace. You come to a place where one path crosses another, and you stand, hesitant at the crossroads, as the trails diverge into the darkness like the spokes of a mysterious wheel. You hope that one of them will take you home, and when, eventually, it does, the relief is sweet and more sincere than the oath you swear that such a thing will never happen to you again, that you are done with luck and serendipity.

And when the road betrays you? When it dwindles and finally disappears, when the night erases all perspective? What do you do then?

A few years ago, the summer my only child left home for college, I moved from an apartment in New York City, to live alone in a small house at the end of a dirt road in the woods of central Vermont.

The house at the end of the dirt road was a small two-story wooden affair, sun-faded to a soft gray and fronted by a wide deck that was rapidly disappearing beneath a feral, unidentified vine. It sat surrounded by thick woods that sharply descended in back to a narrow rocky stream before rising steeply again on the other side, as if in a hurry to get to the next ridge. A secluded, isolated place: at night the darkness and the silence sometimes seemed to swallow it whole.

The inside of the house was a work in progress. There was no flooring except for the original pine two-by-fours laid down when the house was framed, and a slight echo underscored the emptiness of the place. Most of the rooms had not been painted and were covered in a haphazard layer of whitewash. Power to the house was supplied "off grid," which meant that it was not connected to the local electrical network, and relied instead for light and heat on a less-than-adequate array of antique solar panels and a doubtful generator. The wiring was, to say the least, eccentric, and the amount of power available somewhat arbitrary, dependent as it was on whim and weather.

It would be nice to say that the charm of the place was so palpable that it allowed you to overlook all of its manifold faults, but in truth it was a rather plain, stolid-looking house, one that dared you to dress it up in any dreams and promised to teach you a thing or two if you tried.

The road leading up to the house was equally idiosyncratic. It branched off from a somewhat more substantial, better-tended secondary road without bothering to announce itself with anything so assertive as a signpost, and ran rapidly up-hill, paralleling the same rocky little brook. Deeply rutted and pocked with boulders, the road bucked and heaved its way in winding roller-coaster fashion for about a half mile. Then it tipped its hat to a rudimentary driveway before plunging into the woods and narrowing into a bridle path bisected by fallen trees.

A difficult road, barely negotiable and going essentially nowhere, leading to a difficult house, the kind of house that might make sense if it had been left to you by a maiden aunt, but which, when considered as a property you would actually choose to spend money on, looks like a gargantuan mis-judgment. I told myself that I had chosen the house because it was the only thing I could afford. That was true, as far as it went, but it wasn't the reason I decided to live there. I chose the house because of its warts, not in spite of them, because the house's cranky unfinished state reflected my own. One life was over and another was beginning, and I was no lon-ger any of the things I had been, no longer young and not yet old, and because I had to figure out everything all over again, everything—from where to live to how to dress and whom (or even whether) to love, because I had no idea of what to do next, and the middle of the woods seemed the best place to get one. I thought that I would see things more clearly from a place that had no part in my past, the way you climb a tree to get a perspective on the surrounding terrain, to put a name to the strange country into which you have wandered.

I moved to the house at the end of the road to make a new home, a new life, and it was only later that I would see that I had gone to ground, the way an animal does, because I was wounded and beaten and in need of retreat.

I had lived in the apartment I was leaving for twenty-three years. All of my married life had taken place there. It was the warm hearth to which I had brought the work I had done, the newborn infant in my arms, and the friends I had loved; it was the beating heart of my life as a young woman, a wife, and a mother.

I left because all of that had changed.

Time lets you down easily, most of the time—the day fades, the child grows. You trace the last of the light, you tell a bed-time story, not knowing that it is the last bedtime story, the last of a certain kind of homely light, until the moment when it ends, just like that, with the banal shock of a door slamming, with the abrupt lack of ceremony of a tumble down the stairs. The day that same child leaves home is one of these, and de-spite your best efforts, it can knock you off your pins. At least it knocked me off mine.

What had happened was this: I fell out of my own map. It's an easy thing to do, especially in middle age, but really it can happen at any time. We all live by different lights—success, for some, desire for others—and take our bearings along different dreams. Some of us fly west with the night, into the unknown, urged on by adventure; others look only for the harbor lights, and stay safely in sight of home. But whichever way we choose, we come to rely on the sameness of our days, on the fact that

for years at a time the road ahead looks much like the road behind, the horizon clear, the obstacles negotiable. And yet from
time to time we stumble into wilderness. It can happen to anyone, at any age: the graduate putting away the cap and gown,
the fifty-five-year-old rereading the layoff notice, the wife staring at the empty side of the still-warm bed. Now what? they
whisper as they look ahead to a place where the landmarks disappear, and the map reads only TERRA INCOGNITA.

My daughter's departure left questions, big questions, that
her presence and the warm hive of family life had made it easy
to ignore, of who to be and how to live, of what, if anything, I
wanted. By then, widowhood had shaded into a seemingly permanent solitude, without my having thought too much about
it. Similarly with work: I had become a journalist straight out
of college because a close friend offered me a job on a paper
that didn't yet exist, which had saved me the trouble of deciding what to do for a living. Deadline stories written in a news-
room had led to magazine pieces written at home, at a more
leisurely pace: Is that what I wanted now?

I was forty-four when my husband died and fifty-six when
my daughter entered college. I was getting old, and I didn't
know how to do that. So many people seemed to do it badly,
and yet every once in a while, I would see something in the
eyes of an old woman that intrigued me—a kind of triumph,
a knowingness. I wanted to know where that look came from.
I wanted to gather the tools that would enable me to grow old
with grace.

I couldn't figure out any of this in New York. I wasn't sure
why. I still loved the city, but I had lost my place in its furious
striving. I had been a wife there, and a mother, but I had never

been alone and never so confused as to what should happen next. I didn't know what to reach for, and New York is a dangerous place to live in without dreams, no matter how threadbare the old ones have become.

So I went away, to live alone in a place where I knew no one, a place surrounded by a forest in which one could wander for hours without seeing another soul. It wasn't a rational decision; it wasn't even a decision, but an instinct, a drive that existed just below the surface of thoughtful planning. I needed to do this, and I needed to do it without getting lost.

That part was important. I had the idea, at this late and panicky point in my life, that if I could learn to find my way in the woods, then I would find my way through this next part of life. The wilderness is stagnant with metaphor from Dante to Hawthorne to Sondheim, but I wasn't thinking metaphorically: what I wanted, in the most literal way, was a sense of direction.

I get lost easily, always have. On winding country roads, on city streets, in parking lots. On mountain trails so well trodden and clearly marked they might as well be highways—but then I get lost on highways, too, aiming for Albany and ending up near Boston.

For a long time I thought my lack of orientation was merely genetic, like blue eyes or a predilection for the bottle. Usually I didn't mind, and sometimes I liked it—getting lost often led to unexpected adventure. But now that time had begun to thin, no matter how carefully I tried to hoard it, getting lost held less charm. If I were to make my own way in the country of the old, I needed to trust myself, and to do that, it seemed essential that I pare away all that was inauthentic from that mysterious being, my essential self. This scared me not a little: after all these years, I wasn't sure there was much there beyond a large capacity for self-doubt and self-delusion, and a certain agility in swinging between the two.

Experts in the field of direction talk about the difference between way keeping, which is simply the ability to stick to a certain path, following well-marked landmarks and signposts, and wayfinding, what you do when you must rely on yourself, your reading of the landscape and the decisions only you can make. We start out in life learning the first; with luck we end up knowing something of the latter, to the extent that accident and blessing give us a choice. Perhaps in the end that is what wayfinding amounts to: learning how to allow for accident, and make way for blessing.

Scarecrow

Are we lost daddy I arsked tenderly. Shut up he explained.

—RING LARDNER, *The Young Immigrants*

People behave oddly when they're lost. They run around in circles. They backtrack to the last intersection and then find another way to go wrong. They succumb to something called "wood shock," in which they walk around in a trance for hours, often right past the people looking for them and then flee their rescuers. They climb trees, and fall out of them. They follow rivers downstream and end up in bug-infested swamps.

My personal favorite method for finding your way when lost is one that pops up in Canadian survival manuals. You turn around, stopping at four ninety-degree intervals, with your arm stretched out toward the horizon. You have a friend test the relative strength of your outstretched arm in each direction until you find the one where your arm is strongest: in that direction lies the place of your birth. Say you are standing in Nova Scotia and you were born in Pittsburgh and you know that Pittsburgh is southwest of Nova Scotia. With that bit of information, you can deduce the relative positions of north, west, south, and east and so, presumably, the way you should be headed.

Lost Person Behavior is the hottest thing in search and rescue these days. The idea is that you can predict where a person is likely to be found based on their psychological profile. A recent book on the subject identified forty-one different types of people who get lost in forty-one different ways, from children aged four to six, to hikers, hunters, the despondent, the dependent, abusers of drugs, and enthusiasts for extreme sports.

So in light of my particular profile these days—middle-aged, confused—perhaps it wasn't all that odd that I was sitting in the middle of a cornfield, asking a scarecrow for directions.

Besides, this wasn't just some ordinary scarecrow, fashioned from a couple of worn-out brooms and an old coat, or one of those cutesy *Wizard of Oz* types, leaking straw from patched-up jeans and red-checked gingham, that pop up as lawn decorations at Halloween. No, this guy was magnificent: over eight feet tall, barrel-chested, probably stuffed with steel wool, from the looks of him. He was dressed in a pair of old Carhartt work pants and heavy-duty Muck boots, with a gray flannel shirt tucked into thick oil-stained work gloves, one of which gripped an enormous coil of black rubber hose slung over his left shoulder, while the other clutched a pitchfork. A battered old fedora sat low over his face, a bandit scarf pulled up to meet it. A scarecrow that looked as he was meant to look: ready to inflict a terrifying doom on any avian creature dumb enough to challenge his authority.

The woman weeping at his feet conveyed a somewhat less imposing message.

It had been the scarecrow's posture that had caught my attention as I drove by. He was so artfully positioned in the cornfield that he seemed to be striding along at a terrifying pace, ready to mow down anything in his path. This was a scarecrow who knew where he was going. Which was more than I could say for myself, I thought as I studied him.

I was lost. I knew that I was somewhere between Maine and Vermont, but I wasn't sure of much more than that. Normally I would consider such a misadventure fairly routine by my standards, tiresome at worst and a pleasant if unplanned interlude at best. Normally I would keep driving until I found a cell signal, punched up MapQuest on the smartphone, and followed its directions while enjoying the chance to listen to a few more chapters of whatever murder mystery I had going. And that's what I would have done this time if I hadn't freighted this attempt to get from one place to another with a groaning board of metaphysical imagery and symbolic weight, a metaphor for the entire future. Under these circumstances, a murder mystery just wasn't going to cut it, and an empty cornfield in the middle of nowhere became the perfect place in which to unload an entire wagonload of pent-up grief.

Five hours earlier, I had left my only child at the door of her new life, her first semester of college. The last good-bye had been lame; the official ones usually are. Zoë and I had stood facing each other at the edge of an oak-studded lawn, surrounded by the old brick buildings of a New England campus bathed in a lambent late-summer light. We both startled when the chapel bell rang with a proprietary insistence, calling her class to order. I took a long last look at her, as if I could take that image and curl my fingers around it for comfort like the pebbles I used to find in my coat pocket, the ones she used to put there for safekeeping. Finally, with an awkward hug and an apologetic smile, she turned and walked away.

I meant to get behind the wheel of the Jeep and drive off immediately, but when I looked up, I found I could still see her, framed by the windshield, a few hundred feet away. She was standing in a small circle of young strangers, her fellow classmates, listening to the no doubt cheery encouragement of her dormitory's resident upperclassman.

She didn't see me. It was one of those rare moments when I could watch her unobserved, and try to see the person others saw, and not the being that love and familiarity rendered almost invisible. And so I looked at her: a tall, thin, knock-kneed girl, standing in a studied slouch, listing slightly to port with an inherent poise. I tried to imagine her four years from that moment, when I would come for her graduation, what she would be like then, how she would have changed. But I succeeded only in remembering the wild griefs and mistakes of my own college life and turned away quickly. It is a tricky business sometimes, to see your child as she is, to let her step out from behind the scrim of your own mistakes and regrets, your fierce and futile hopes for her own unmarred happiness.

Finally the small group turned and shuffled away across the lawn. I bent my head to the road map in my lap. Woodstock, Vermont, was about two hundred miles due west, more or less, of my present location. The officially sanctioned route, as I thought of it, thereby investing it with an authority to which it had never laid claim, the one recommended by Google Maps and MapQuest, was simple, if counterintuitive: head south on U.S. Interstate Route 95 until Portsmouth, New Hampshire, then west across the state on Route 101, and then finally north on Interstates 93 and 89 until the first exit in Vermont, which would drop me about twenty miles away from what was supposed to now be home.

This was not only the route my cell phone liked, but it was also the route recommended by that ultimate authority, the guys at the gas station. Two gas stations, in fact, the Sunoco in Woodstock, Vermont, and the Texaco in Brunswick, Maine. Both caucuses had pulled on their polite, patient "we're dealing with a flatlander" faces when asked about alternative ways to get from one place to the other. Sure, there were other ways to get there, they said. But they're slower. They uttered the word with emphasis and a kind of creepy finality, as if by "slower" what they really meant was "filled with things furred and fanged, lit by glimmering swamp light and haunted by the wraiths of vengeful women hunting down their faithless lovers."

But that afternoon, I wanted no part of superhighways relentless as time, speeding out of one map and into another. I needed those slow-poky little roads. They would be dotted with small towns and country stores and old barns, farm stands, and intersections that invited you to stop and look around, idiosyncratic places that could bolster the reassuring sense that life continued, despite the fear beginning to hammer at the back of my brain that mine had hit a wall. These roads would take their time, in a syncopated rhythm of hills and flats, orchards and pastures, villages and strip malls. They were old routes shaped by the landscape, not blasting through it, and the places that grew alongside them were testimonies in brick and wood and marble to what had mattered to the people who lived there. In their details, I would see how Maine differed from New Hampshire, and New Hampshire from Vermont. I would be passing through somewhere, rather than anywhere.

I studied the map. That is to say, maps: I had a book of Vermont maps, and another of New Hampshire maps. (What I didn't have was a GPS. That was cheating.) Both books chopped up each state into squares, one square to a page, none of them necessarily contiguous. Flipping back and forth among the grids, I figured out an alternative route that looked more direct and eschewed all highways. It was composed of a somewhat bewildering tangle of what were probably two-lane blacktops that unspooled like a web of thin spidery veins over a half-dozen unconnected pages. It would be a little tricky in places—New Hampshire in particular seemed dauntingly chockablock with massive lakes and mountains. But the beginning part was simple enough. I identified a smallish-looking state road that seemed to split off from Route 1, the Maine coastal road, at about the right latitude (if that was even the word I wanted), although it seemed to disappear once it arrived in New Hampshire. No matter, there was another one going in more or less the same direction. A lot of them, in fact. They crissed and crossed and do-si-doed all over the place, but eventually some of them evidently intersected with Interstate Route 4, which would take me directly into Woodstock. There were also a lot of alternative, even thinner lines that might or might not provide a shortcut, and a few roads that seemed to change their names simply for the hell of it before popping up in unexpected places. I thought about writing down a route but decided against it. It looked simple enough. And I wanted an adventure.

This day, after all, was not only an end but also a beginning—of a new life in a new world. I needed to get to where I was going according to my own lights, along a path I had chosen, not one generated by some witless computer program, or traced out by helpful strangers.

Later I would learn about route delusion and disorientation behavior and a whole lot of terms scientists use to characterize the bizarre ways people make an utter hash of getting from one place to another. At the time, however, I tossed the maps onto the passenger seat and pulled out of the parking lot, blinking away the last of the tears. You'll be fine, wiser heads had told me when I asked them how they got through this day. You'll cry, but then you'll feel light in a way you haven't felt in years.

I had doubted that second part, but as I slipped out of Brunswick onto the highway that would in turn tip me onto the coast road, I sensed it, that first stirring of exhilaration. It was a fine day, there was plenty of it left, and I was headed into a brand-new life, one in which, for the first time in many years, I had no idea what would happen next.

My husband, Lee, had died when Zoë was six, and in the beginning, and for many years afterward, life was a matter of putting one foot in front of the other. I don't mean to say there wasn't a rich complement of joy and deep satisfaction, along with the usual hardships of single parenthood, to accompany the archipelago that grief requires us to navigate, only that the way was straight and the direction clear. I had a child to raise and a living to make and I did these things cocooned in the warmly reassuring diurnal rhythm of playdates and sleepovers, of tête-à-têtes with other mothers, of teacher conferences and bake sales—and with little concern over the parts of life that might be missing.

But after my daughter entered high school, the soft focus and narrow spectrum through which I was used to viewing the world shifted sharply. Her new school politely made it clear to the freshmen parents that their involvement in their children's academic careers was essentially limited to attending school events and writing large checks on a regular basis. My daughter passed into that seemingly endless phase of adolescence during which an adult's presence was mainly required at times of intense unhappiness. More and more vacations and weekends were spent at other people's country homes or on the kind of educational jaunts meant to impress the jaded eyes of the college admissions committees in the rapidly advancing future. Soon, too soon, she would be gone.

I had begun to sketch out the outlines of my response to that inevitability about ten years before, though I didn't know it then. One of my two stepdaughters had been married at her mother's house in Barnard, a small hamlet about ten miles north of Woodstock. I was struck by the beauty of the place, a country of green hills and old barns and grazing cows, and by the faint breeze of memory it stirred of childhood visits in summer and at Christmas to the hardscrabble speck of a town in southern New Hampshire where my father had grown up.

The decision to spend time in Woodstock alone was a last-minute idea, impelled by my daughter's going to summer camp for the first time, by a July heat wave that made the prospect of an air conditioner–less loft in New York City unappetizing at best, and by the startling realization that for the first time in a very long time, I was facing a month in which I could do as I pleased.

Parents lead contingent lives, the personal put on hold, decisions of what to do next based on the needs, well-being, and schedules of others. That summer when it changed—if only for a month—was, I see now, a small intimation of what lay waiting around the wide bend of my daughter's childhood: the beginning of the rest of life.

The house I rented, sight unseen, and selected from the scant few that were still available, reflected the extemporaneous nature of that decision: it was an odd little place, painted a violent shade of pink, which perched precariously on a narrow bit of ground above the wide shallow river that wound its way through the village.

My temporary roost sported an eclectic conglomeration of architectural elements and was only sparsely furnished, the result of the recent divorce of the house's owners. The bareness of the place appealed to me—as did, to a much more limited extent, the ex-husband. Angus was a Ph.D. in botany who worked nights as a waiter so that he could dedicate his days to . . . well, no one was really sure how he spent his days, beyond Rollerblading and executing ill-conceived, highly public pranks, usually performed in costume and received with irritated wonder by the village's inhabitants. But he was a connoisseur of the country's beauty: he once took me to a waterfall hidden deep in the forest, a place so lovely that you could almost believe that dryads and fairies were real, I observed. Angus looked at me with a faintly pitying surprise. "Of course they are," he said.

For the first time in years my hours were for hire, freelance agents ready for any employment. Zoë's summer camp was one of those shoestring operations where the 1960s lingered on, sustained by large doses of Joni Mitchell and politically correct entertainments like the daylong reenactment of the Cherokee Trail of Tears, which was offered up under a broiling sun for Parents' Visiting Day ceremonies. No cell phones or computers were permitted, which meant that Zoë and I were unbuckled from each other for the first time, apart from the handwritten letters detailing cold morning swims, mean girls, kind counselors, and an aggressive mold population that was slowly turning everything in her tent, including the inhabitants, an alarming shade of gray.

Every morning I walked down the steep hill and bought coffee and a pastry from the somewhat self-consciously European but extremely good café and then trudged back up again, amazed at the spaciousness of the day ahead. In the beginning I took the coffee and the pastry up the winding stairs to the odd little crow's nest of an office at the top of the house, where I tried to make headway on a book I was supposed to be writing, although most of the time I just watched the progress of the indefatigable wasp who was patiently attempting, so far unsuccessfully, to wedge himself through a gap in the screen that covered the window. In the afternoons I would calm my anxiety over the work I didn't do by climbing Mount Tom, a gentle old hill laced on its southern side by an easy switchback that used to bring turn-of-the-century ladies and gentlemen up to the summit for the view. Or I would walk along the Ottauquechee River and up into the hills on rolling country roads, hypnotized by the green buzzing beauty of the place.

In Woodstock I felt lighter, at least when I wasn't trying to work. Here was a place where I was none of the things I had been, not widow nor wife, not mother, a place where I would not encounter the ghosts of old selves and old lives. A place where I could imagine being a woman alone in a future still safely far away.

One morning I walked past a woman hunkered down over some window pots outside the Yankee Bookshop, planting orange and scarlet dahlias. She was blond, buxom, and barefoot, in her forties, presenting a round face full of frank curiosity and a glint of mischief in her bright blue eyes. Her name was Susan Morgan, and she was the owner of the store.

She looked me up and down. You can't be on vacation, she said. You look terrible. The snaky dread began to fill my lungs. I'm writing, I said. Or not writing. Mostly I'm having a little bit of a breakdown. It's kind of a full-time job. Then I stopped. In New York, people tended to discuss breakdowns and anxiety attacks with the same casualness they did the difficulty of finding a cab during the afternoon shift change. But Woodstock wore an air of such self-complacency that the comment felt like very bad manners. Susan looked at me thoughtfully and asked a few questions about where and with whom I was living. I told her. Well, that's the problem, she said. You're too alone. Of course you're going crazy. Why don't you write here?

Normally, I would have thanked her kindly and moved on—I had a complicated ritual when it came to writing that stopped just short of human sacrifice; it was not easily transported. But I was desperate, and the project was long overdue. The next day I brought my laptop to the bookstore and Susan set me up on a couple of billowy floor cushions in the travel section. I wrote two chapters.

That was the first of many visits to Woodstock. A month later, the World Trade Center fell and the night sky was lit by two beams of blue light marking their place and the way in which the future had changed forever. Woodstock, its beauty, its safety, took on a talismanic quality; I returned every summer while my daughter was in camp. I came to know a few people a little, the way you do when you see them on the street while doing errands, or behind the counters of the stores you frequent. Their warmth in welcoming me back each successive summer was gratifying. And though I barely knew them, I wanted to be one of them, no longer an outsider like all the other visitors with their impatience and their restlessness.

I found a real estate agent, Lynne Bertram, a woman whose own roots ran deep in the area—she was the daughter of Wallace "Bunny" Bertram, who had operated the country's first mechanical tow for skiers, fashioned out of 1,800 feet of rope and a Ford Model T up on Gilbert's Hill just outside of town. When she wasn't selling real estate, Lynne helped her husband, Nelson, to run a two-hundred-acre farm, where they raised Scottish Highland cattle, enormous beasts with great shaggy heads and arcing horns, and a small but ever-growing flock of sheep, whose lambs she named after French cabaret singers. In her spare time she presided over an orchard of ancient apple trees and a spectacular flower garden. She was a handsome woman and had been a great beauty and a professional ski racer on the international circuit in her youth. She was not one to mince words.

We sat at the counter in the coffee shop, where Lynne reminisced with her cousin Sandy, the café's manager—Lynne had been the flower girl at the wedding of Sandy's aunt, and they talked about the dresses the grandmothers had made for the bridal party—and I envied the way in which they were stitched into each other's lives.

We ate grilled cheese sandwiches and I told Lynne what I wanted. A small windfall from a real estate sale had made it possible for me to buy a house out in the country—nothing fancy. I'll see what I can do, she said, and eventually she found me the house at the end of the road. I hadn't spent much time there while Zoë was in high school, and when I did go I went alone. Zoë had come with me once, her freshman year, but she had a city kid's allergy to the country: she couldn't sleep; it was too quiet. So the house became my retreat, the place where I could slip away when the noise of life began to overwhelm. I thought of it, a little sheepishly, as my Fortress of Solitude, like Superman's Arctic redoubt in the comics. When I was a girl, whenever family life became too intrusive (which, of course, was nearly always), I would lie in bed and read about his trips there and envy him desperately. One day, I had promised myself, I would have such a place of my own. And now I did.

In those days, I visited the house like a shy courting lover, stealing a weekend here and there, or a week of Zoë's high school spring break. I brought flowers and books and other votive offerings, and because I was there so rarely, the place remained a Shangri-la, both intimate and exotic. I was entranced, like many a refugee from the city before me, by the seeming simplicity of my life there, the quiet order of the hours. Because I depended so much on the sun for power, my days followed a simple progression: I would get up just before dawn and watch the first of the light cast a leafy shadow play across the living room wall while I drank my tea. The window by which I sat overlooked a small meadow that gleamed gold and green in the early-morning light; sometimes, I would linger there, hoping for a deer or bear or moose to wander by. By the late afternoon, I was upstairs in the rocking chair in my bedroom, where I could take advantage of the last of the western light while I read or knitted and dreamed of a life where such ordinary pleasures and welcome constraints could be the rule rather than the exception. And now they were about to be.

Coastal Route 1 ambled past Portland, where I picked up Route 25, which seemed to head west in approximately the right place. The late-August light slanting across the road had the valedictory quality of late summer, and I wasn't sorry to see the season come to an end. I wanted autumn to arrive quickly; I want everything to be over. I was so sick of good-byes.

The spring had been endless. Zoë had been admitted early decision to Bowdoin College, and the whirlwind of anxiety and deadlines in which we had lived from the beginning of her last year in high school had abruptly given way to the torpor of the second-semester senior. I discovered with some wonder how much real estate in my brain had been taken up over the last three years by deadlines and the ratios of applications to admissions, by grade point averages and achievement scores, and by the emergency consolations required by midnight meltdowns over the entire spectrum of late-adolescent angst—from the fatal mediocrity that would result in her rejection by even the stoner college that was her safety school, to the dismal number of tags and pokes, whatever they were, that determined her popularity on Facebook, and therefore her self-worth. Now the days were devoid of most of these woes, and I realized how little thought I'd given to what was to replace them.

The school counselors talked about a phenomenon called "fouling the nest," in which the child of your bosom begins to act like the scorpion in your shoe, an adolescent's defense mechanism, apparently, to make it easier to leave home. I would look at the ceiling and count the days until her departure, thrilled not to be dreading it anymore, on the rare occasions of her bad behavior. But then something would happen, Zoë would play a song that had been her favorite when she was twelve, or her two best friends would saunter in, with a strange new self-possession that had come in the mail with the acceptance letters, and I would remember the night I had nursed one of them through her first wretched encounter with vodka, or some silly afternoon I had been lucky to be a part of. And I wondered if I would ever know the women they were on the verge of becoming, now that the road had opened and they were setting out in so many different directions. The memories skittered like dry leaves along the bare wooden floors of the apartment, echoing and reechoing, and then swirling away.

Around this time, I went to a book party with an old friend. It was held in one of those beautiful old West Village apartments, crowded with a worn velvet sofa, easy chairs, and a settee draped in quilts and vibrant silk cushions. Thin spring sunlight filtered in through a big bow window overlooking Washington Square Park. A gray cat dozed amid the chatter of perhaps a dozen guests, worldly, successful people from a variety of walks of life. There was a grave, bespectacled Turkish photographer, a celebrated biographer, a stylish editor from a well-respected publishing house, and a plump grandmotherly woman who had known the author since she was a child.

The guest of honor was a woman about my age, whose sixth novel we were celebrating. She was married to an old college classmate of mine; they had six children. She was a professor at an English university and the official American translator for an eminent novelist. She was dressed in a black dress and black stockings that should have looked dowdy but instead came together in the kind of elegantly careless disarray that is somehow the hallmark of confident British intellectual women of a certain age.

We were introduced and she was very kind, but when she asked me about myself, I could think of nothing to say. I mumbled something about being at a turning point, about my child going off to college, and trailed off midsentence, unable to remember what I did with my days.

I walked back home, shaken. This woman had done more with her life in one month than I had done in years. To witness her world, the assurance with which she seemed to move through her life, was to recognize the distinction between a real person and a ghost. I tried to muster a more accurate image of my own life, the work I had done, the friends I cared about, but my confidence was fading fast, and as I walked along, the city in which I'd lived for so long felt cold and alien, its endless variety providing no solace. Everyone looked strange—the doughy, exhausted guy in the coffee shop changing a soggy coffee filter for the millionth time, the hard-edged scramblers charging down the street barking into their cell phones, even the people who usually made me smile, like the compact young woman dressed in four shades of turquoise, neat as a pony in her high heels, rushing along on the stream of her own vitality.

I came home to find Zoë standing in the kitchen, dressed up for a night out in an old crocheted dress with torn black stockings and hand-me-down spectators, staring disconsolately at a tub of frozen chili. I had barely seen her for the last few evenings, but that night her plans had changed at the last minute and she was home alone. She couldn't get the container open. The microwave wouldn't work. She was hungry and close to the tears too little sleep and a surfeit of emotion bring. I smiled, happy to have stumbled into a temporary reprieve. I made her dinner, and we sat down together on the old blue sofa, watching animated movies from her childhood late into the night, counting the last moments of her childhood as they ticked away.

I began to make plans and lists and more plans, and lists of plans, charms against the uneasiness conjured by the book party and Zoë's looming departure. I would move to Vermont, to the little house I had bought. I would buy a dog and live in the country. I would reinvent myself, a woman alone, solitary and self-contained.

I trolled the Internet, in search of a car that could make it up my cranky dirt road and a puppy easy enough to manage with my paltry lack of disciplinary skills. Gradually the rough draft of a new self emerged. I saw myself striding through the countryside, needing no companionship but the noble hound who followed close to heel, the evening a cozy pastiche of firelight, softly playing jazz and great books. At times the vision was so real, so beguiling, that I was impatient to be off; I couldn't wait for the beginning of this perfect new life. I navigated the streets of the city with an already nostalgic fondness.

Zoë was skeptical. *You're going to live alone, in the middle of nowhere, in the country? Besides*, she said, *what will you* wear? Flannel? She liked the idea of the yellow Lab puppy I eventually contracted to buy, or did until she discussed it with her friends. They say you're getting another little blond baby to replace me, she reported with some indignation.

If the spring seemed at times to never end, the last weeks of her senior year cascaded by in a torrent of ceremonial moments savored, a looking back that was finally divested of sorrow and mourning. It began with a graduation party, at the house of an old friend, one of the first people I had met when Lee and I moved to New York: We had been pregnant together, and our daughters had been born six weeks apart. Now we stood in the fading warmth of a spring evening, looking at each other and at the parents we had known over the years, fellow travelers through the rapids of childhood, men and women who were grappling with the same images, light and dark, all of us runners at the end of the race, and I wanted to lift a glass to each and every one of them and say, Well done.

And then the sight of Zoë in her prom dress, hovering at the door, with a look on her face that asked, How do I look? Not in the casual, is-there-anything-amiss way of every day, but the question of a girl transformed, wondering, has the miracle occurred? A little abashed, for the first time in a long time, nervous, expectant, hopeful. And yes, yes she was, radiant and beautiful, a young girl alight, kindled by the excitement that only can come to one so young.

She came home late the next morning straight from the after party, wobbling along in SpongeBob SquarePants pajama shorts and her high-heeled silver sandals, drifting happily in the afterglow of a night that had been everything she had hoped, such a rare thing, even in childhood.

She left for a trip to Europe a few days after that. Crossing the street to the grocery store, I was reminded of how I'd felt the first time I crossed the street after she was born, the care I'd taken, awestruck at how much it suddenly mattered that I take such care. And now?

The shadow the scarecrow cast was lengthening. The birds had long since departed. I was somewhere west of Maine and east of Vermont, but that was about all I was sure of. New England was apparently parsimonious when it came to road signs, and the ones I did encounter were confusing. Did the right-hand-turn sign indicate the road immediately in front of it or the one just slightly beyond? Why would a road running east to west give you the option of going north or south?

Maine had been straightforward enough. I had headed west on Route 25, ambling past small towns and villages, Gorham and Standish and Cornish, admiring the staunch English names, so different from the suburbs in which I had grown up, the Camelots and Mantuas—fanciful, almost poignant attempts at imposing history and romance on the raw and recently constructed.

But in New Hampshire the road I was following betrayed me, disappearing off the map altogether before popping up quite a while later, dressed in a different color (did that mean something?) and going off in what looked to be a nearly circular direction. In between I wandered around in a chaos of wrong turns and misunderstood directions from the few living souls I encountered; at one point I found myself on the outskirts of the White Mountains. I contemplated cutting through them, a plan that failed due to the lack of an actual road, before threading my way for hours around the fingers and inlets of pine-shrouded lakes until I emerged somewhere a good deal north of where I was meant to be.

There was more—the wrong interstate in the right direction, the right interstate in the wrong direction, not to mention simple road signs transformed into opaque Zen koans by frustration and fatigue—until I reached the point that I wanted only to find a motel room with a working TV where I could hide for the night. To that end I exited whatever interstate I was on, numbly driving past wide, thickly planted cornfields, stripped of their yield and beginning to yellow.

That's when I found the scarecrow, stopped to take a picture, and ended up sitting on the side of the road, wondering how I could be so hapless. Finally, I got back in the car and headed off in the direction in which my guide was striding. That road, miraculously enough, took me to the road that got me to Route 4, the way into Woodstock.

An hour or so later I started up the hill to my house in a dusky darkness, the remaining light from the setting sun obscured by the overhanging branches of the maple trees. Three-quarters of the way up, in the middle of the road, stood a large object, its shape obscured by the shadows. As I crested the rise, however, an errant patch of sunlight illuminated the form just as it unfolded its enormous wings and, with a great beating motion that stirred the dust at its feet, rose silently into the air.

Earlier that day, caught up in my tremulous worries and excited visions of the future, the sight of a great blue heron on an empty country road would have struck me as a sign, a message from the future that all would be well. But I was too tired and too humbled for any such egocentric nonsense now.

I was simply grateful for its beauty, the unexpected grace of its presence. I gunned the car over the last hill and turned into the driveway.

The house was almost pretty in the twilight. The overgrown grass was alive with wildflowers, and the mysterious vine that was choking the porch railing and encroaching on the front door was a brilliant green; it looked lush and welcoming against the soft gray of the shingles. To my delight there was an apple blushing on the scraggly tree at the top of the drive. Inside, I knew, the bare rooms would be dark, and there was an odds-on chance the lights would not be working. There would be no food, and probably no sheets on the bed. So I stayed outside for a moment longer, tracing the smooth curve of the red apple ripening on a tree that stood in front of the house that might be home.

2

The University of Guam

I am mainly ignorant of what place this is.

—WILLIAM SHAKESPEARE, *King Lear*

T he morning after I arrived, I made a cup of tea and waited for the familiar sense of contentment to appear, the way it always had done the first morning of my visits. But you look at a place differently when you plan to live in it, no matter how often you visited. A trill of unwelcome unease surfaced.

The house that was to be my Fortress of Solitude was located in a part of Woodstock that bore little resemblance to the area with which I was familiar from my August rentals. After that first summer in the pink house in the village, I had stayed in a succession of converted barns and carriage houses in the surrounding hamlets to the east and north, all of them located on good roads in spectacularly beautiful countryside, a panorama of rolling, manicured green hills punctuated by occasional flocks of sheep or a small group of black-and-white cows gathered picturesquely near an old stone wall, and crested by sun-dappled stands of timeworn trees.

Such beauty came at a price: the upper Connecticut Valley, in which Woodstock is situated, comprises some of the most expensive real estate in the state, and the surrounding areas of Barnard and Pomfret and Bridgewater are home to the extravagantly wealthy.

I had almost despaired of finding anything I could afford when, at the end of a long day of driving about with Lynne Bertram and her two yellow Labs, Lyla and Sandi, she hesitated. There was a place a little farther afield, she said. It was remote, it was off the grid, the road was bad, she hadn't seen it, but yes, it was affordable.

Let's go, I said. Later, the frantic need I felt then to claim some part of this unfamiliar place would seem inexplicable to me—after all, the only thing I really knew about Vermont was that it was not New York. But perhaps, at the time, that was all it had to be.

We drove south from the village on Route 106. Most of the cleared land to the left belonged to the Green Mountain Horse Association—the area was the birthplace of the Morgan breed of Thoroughbreds—and the barns and corrals, the steeplechase courses and parking lots for the horse vans, bespoke a world of wealth and obsession that sheltered in the surrounding hills. On the right a few more modest houses were strung out along the shoulder of the road—a stone cottage, a faded red barn, a 1960s-era chalet-style A-frame, a thinly disguised trailer—followed by tree-shrouded lanes that swung away discreetly up and out of view before they could permit a glimpse of the larger estates.

We passed through the village of South Woodstock—a cluster of buildings that included a general store and the post office, and the stately Kedron Valley Inn, where generations of shiny blond ponytails and polished black riding boots had put up during the summer show season—and continued on up the road past a heavily wooded stretch of land, until we reached a wide-aproned turnoff to the right. We bumped along for a half a mile until the road intersected a rocky creek and a much narrower unnamed dirt road diverged to the right, wedged between the steeply rising ridge to one side and the creek on the other.

It was March, and mud season had rendered the road little more than a boulder-strewn streambed. Lynne's SUV was having a tough time making it up the steep climb, the left side of which was sheathed in ice, while the right was a grid of waterlogged furrows. Lynne wanted to turn around, but by then I had caught a glimpse of white shutters and a curl of smoke rising above a screen of maple trees. Within the hour I had seen the house, and that summer it was mine.

The house was twelve years old then and had been built nearly single-handedly by a retired engineer and his wife, Bob and Tess Riley. The Rileys were in their late sixties when they moved to Vermont from Maryland. Initially they had simply planned to invest their limited savings in a larger house in a less expensive place. Instead they bought a ten-acre patch of hillside belonging to four hunters, whose sole improvement to the place had been the construction of a slapdash deer hunting stand high up in one of the larger maples. There was no water, no septic field, no electricity, and no road bigger than a footpath.

Riley himself knew nothing about building houses, apart from some rudimentary plumbing and carpentry skills learned at his father's knee. He and his brother walked the woods until they found a flat part large enough to lay a foundation; they cleared enough room for a house and garden with a single chain saw. They moved in when there was nothing more than bare walls, a rough plank floor, and a five-gallon plastic bucket for a toilet.

For the next ten years the Rileys hammered sawed plumbed wired jerry-rigged and improvised their way into a home of their own. The result was an idiosyncratic marvel. The ground floor was laid out according to an open plan, with half of the floor on a level three steps higher than the other. The steps themselves were bisected at a ninety-degree angle by the staircase, necessitating a tricky little dance movement to get from one set of steps to the other. A short hike from the kitchen, on the far side of the mudroom, a kind of pantry lay in wait, dominated by an enormous gas-powered chest freezer, like the ones on TV that tend to fill up with dead bodies. The pantry adjoined the garage, which had been truncated and angled away from the driveway to make room for a plastic greenhouse that had enclosed Tess's orchid-raising operation.

The house was still unfinished when the Rileys decided to sell—they had no wish to put it on the market, but they were out of funds, and Tess had developed emphysema; she could no longer walk upstairs to their bedroom or down the hill to the creek that had first charmed them into buying the place. After the sale, the Rileys moved to the neighboring town. Tess died five years later, a difficult, angry woman to those who didn't know her, famous in the neighborhood for rushing out of the house to scream at the horseback riders who had the temerity to pass her house on what was in fact a public road. Bob Riley, a tall, thin, quiet man, would fall in love with his next-door neighbor; last I heard, they were looking for a house that was neither his nor hers but theirs.

By the time the Rileys moved out, the house had evolved into a proud testament to one couple's stubborn imaginative vision, but it was also an awkward eccentric dwelling whose deficiencies were tempered by unexpected grace notes. The wiring was a tangled nightmare that hung in loops from the basement ceiling, the garage had no roof beyond the already sagging tar paper and vinyl sheeting, the fireplace lurked in a kind of no-man's-land between the living and dining rooms, the living room was a small forest of wooded columns thrown up wherever it looked like more support was needed. There were doors that led out to nothing but blue sky and stairs where none were needed and walls that never got to where they were meant to be going. But there was charm there as well. Bob Riley had taken the time to carve a delicate motif of undulating vine leaves along the door lintels and the wooden banister, and those small bits of loveliness hummed a kind of tune to me every time I passed them.

When I bought the place I thought I had put aside enough money to fix the most glaring problems, but most of that money had gone to solving a series of mysteries in the plumbing system that had generated astronomical heating bills. I would just have to live with the rest, including Tess's wallpapering choices—giant lurid vegetables (kitchen) and depressed farm animals (mudroom).

It was very quiet that first morning of my permanent residency, overwhelmingly so. The sky was gray with low unbroken clouds, the air heavy and still. It had been a dry summer and the water level in the little brook was too low to make much of a stir. Even the summertime buzz and hum of the insects was hushed. I walked from room to room, which still smelled of sawdust and mouse droppings and dust in the way of uninhabited country houses, and looked out each of the

windows. The house sat in former swampland and, except for
a small meadow to the east that covered the septic field, was
surrounded by steeply rising woodland. Fifty years earlier,
this part of Woodstock had been populated with small family
farms, but the forest had long since devoured most of them,
and the woods outside the windows were dark.

The house was nearly empty, except for a few items the Rileys had sold to me: a bed in one of the rooms upstairs, and
downstairs, a dark green leather La-Z-Boy recliner whose
spectacular ugliness was matched only by the instant discom-
fort experienced by anyone who ventured into its depths. In
the past, I had almost admired the chair's utter refusal to meet
even its basic duties of form and function, but that morning it
was too perfect a synecdoche for the house itself.

I had bought a few pieces of secondhand furniture in New
York before I left; they would be delivered soon, and the mere
thought of them provided some comfort. I had deliberated
carefully over those few cautious purchases. They represented
the first time I had chosen my own furniture. As a young
woman, I had raided my parents' house for castoffs, while
nearly everything in the loft had been handed down from
my husband's family. I had taken nothing from the New York
apartment, because it seemed wrong to loot the place that
was still home to my daughter. Besides, the pieces of polished
chrome and leather from the 1950s reflected a long-vanished
world of New York sophistication that had been Lee's but
never mine. They would look out of place in my new home
and I didn't want them: they were part of what I was leaving.
I had lived too long in the rubble of a life that had exploded
with my husband's death. The new furniture—a wobbly and

scratched-up green nightstand from a secondhand store in the city, a gigantic white painted table from a Kmart in Lebanon, New Hampshire, a lamp with a base of curling leaves and metal roses and an oversize shade of battered parchment, so atrocious it was attractive—wasn't much to look at. But these pieces represented my first attempts at defining what sort of life I wanted to live now. Perhaps when they arrived, these ordinary artifacts would mute the echoes my footsteps made when I walked into the rooms, and still the voice that was beginning to wonder what I was doing alone in this house in the middle of nowhere.

We begin in dreams, tatty, foolish, romantic; we end in ashes, which, if we are lucky, allow us to begin yet again. The carapace that had been my fantasy of life in Vermont would burn off slowly, and I would cling to it as long as I could, because even as I watched it begin to evanesce, I could not imagine what would take its place.

Some part of my brain, in those first few weeks, knew that the wise thing to do would be to simply get back to work. "There is no greater cause of melancholy than idleness, no better cure than busyness," a scolding voice reminded me, quoting from Robert Burton's *Anatomy of Melancholy*. I had proposals for a few magazine stories I needed to write, and notebooks to comb through in search of ideas for longer projects, but the assertion of authority required to unpack the boxes, the conviction that I had anything left of a writer's curiosity and passion, were packed away as well. Besides, these were the days before the stock market crash and the recession, and there was just enough money trickling in on which to get by—just enough, if you factored in my la-di-da attitude toward going into debt. (I wonder, sometimes, if there wasn't a drastic drop in the number of midlife crises taking place in this country once the bottom dropped out of the American economy.)

Since I couldn't seem to focus on work, I made lists. Long lists, of everything I needed to fix, or understand or abandon or acquire. At first there were a lot of short lists, but in the end I boiled them down to two: List Practical, and List Metaphysical, which I kept in a notebook labeled F OF S (for Fortress of Solitude, I'm afraid).

List Practical was very long and covered everything that needed to be done around the house and grounds. The items ranged from Buy Tomato Plant to Deal with Electrical Wiring, to Learn Difference Between Weeds and Flowers, and the list grew like kudzu as each day taught me another thing I didn't know.

List Metaphysical was much shorter:

Get Sense of Direction
Find Authentic Way to Live
Figure Out How to Be Old
Deal with Sex
Learn Latin

This last item might sound as if it should have gone on List Practical, if it went anywhere at all, but I figured if I became the kind of person who ended her day at the kitchen table (when I had a kitchen table) with her head bent over Cicero, then all the other items on the list would have been taken care of and I would probably have some time to kill.

The biggest item on the first list was becoming remotely comfortable with living off the grid, which meant an immersion in a set of responsibilities and chores that were as foreign and exotic to me as the customs of a country whose language I didn't understand.

It would have been daunting if the system worked with clocklike efficiency, but this was not the case. About a year after I had bought the house, I had replaced the balky backup generator that rumbled into noisy life when the sun was shining brightly and it therefore wasn't needed, and retreated into sulky silence when it had been raining for three days straight and there was nothing left in the solar battery to power the lights. The new one was a model of efficiency, but it nevertheless required the same level of care as a three-year-old with a bad cold. A young man named Dave came over to teach me about logging oil levels and working hours and how to switch the thing from automatic to manual in order to maintain the equipment. This, in turn, involved dealing with an enormous colony of wasps whose nest was established in the generator shed. Probably want to get rid of that, Dave advised. At night, he said, when it was cooler, and they were asleep. Or better yet, wait for fall.

The solar power system was even more complicated. Outside, on an alarmingly fragile-looking wooden platform, were the twelve gray solar panels that would have to be swept with a broom whenever it snowed. Inside, in a corner of the basement, lurked a massive bank of solar batteries and the inverter, a bewildering, blinking, flashing, button-festooned electronic box that coordinated the rest of the system as well as its interaction with the generator.

I tried to read the manuals the Rileys had left, but the technical language made my eyes cross. I called around to the local solar energy companies and found someone who agreed to review everything with me. Howie was a tall, lean young man with bushy hair beating a premature retreat from his forehead, the kindly, curious expression of Big Bird, and the calm patience of a budding bodhisattva. He would stand with me for hours in the musty basement, in front of the blinking yellow, green, and red lights, his eyes blinking in response as he listened to my anxious questions and tried—once again—to explain amperes and volts and solar cells and the necessity of bringing the system up to float. I would take furious notes, which, when I read them over again, would consist of numbers and arrows and abbreviations and illustrations of the sun that might as well have been Mayan glyphs for all the sense they made. Howie was sanguine. It's a very steep learning curve, he said. But you'll get there. I think what you're doing is very brave. That last bit worried me a little; I was pretty sure I didn't want to know what he meant by it.

After Howie came Dan, dark-haired, wiry, with a long ponytail threaded with gray—he was growing it long in order to donate the hair to be used in wigs for chemotherapy patients, in memory of his brother, he said, who had died recently and very young. His love and his anguish were written on his face. Dan must have had less confidence in my abilities, or more experience with newbies, because he tried to keep things simple. He showed me how to keep the twenty-four battery cells filled with hydrochloric acid and distilled water, what numbers to watch for on the complicated electronic menus in order to gauge the relative health of the system, and most important, his home phone number in case things went wrong. Which he was pretty sure they would: The batteries were old, and the inverter's software was written in the technological equivalent of pidgin English. Besides which, Dan didn't think I had enough solar panels to power the house full-time, and even if I did, the surrounding woods kept my hours of available sunlight to a minimum.

I would have to be very conservative, he said. Don't use the clothes dryer, let the sun do that job. Don't run any appliances late in the afternoon—use the vacuum cleaner on a sunny day, not a rainy one, so that the cells would have a chance to recharge. Forget about the garage door opener or the device that churned up the water in the bathtub. Don't even think about the dishwasher, and unplug everything that wasn't in use.

I began to grow a little afraid of the house. I would think about the solar cells emptying and the generator choosing that moment to fail and what it would be like in the coming nights, when the house was as black as the night sky and I was alone. I was terrified of waking up in the middle of the night to find I couldn't turn on the bedside lamp. I bought a dozen flashlights and checked and rechecked their locations, but it wasn't terribly reassuring.

Luckily I had a very long list of things I thought I needed, and in the mornings I would flee the house for the safety of my car. I had a wonderful car, a four-door Jeep Wrangler, in no-nonsense army fatigue green. I told myself that I bought the car because it was practical, because it was one of the few vehicles that had a fighting chance of getting up my road in winter and mud season. But the truth was that I am a suburban girl at heart, for whom a car had always signified and would always signify freedom. Driving was meditation and escape and power: driving was bliss. I fell in love with that rig, as I learned to call it. At its wheel I was cool and swaggering and bulletproof and invulnerable; if I could have lived in it, I would have. The chassis was high off the ground, and behind the wheel I was lord of the forest and everything unfamiliar. I would kick off my shoes, turn the music up loud, kiss the accelerator with my calloused foot, and drive way too fast down the bumpy road.

Nothing in Vermont is close to anything else, particularly around Woodstock. Apart from Gillingham's, the wondrous but expensive general store, the town has a genteel horror of the utilitarian; even the two hardware stores are safely outside the village. For most of the practical stuff of life people tend to go farther afield, past the piratical antique stores, and the odd little tourist baits selling Scottish goods and lawn ornaments shaped like cows, and the little bakeries and handmade pottery shops, and the occasional auto body shop to Interstate 89, exit 20, the first in New Hampshire, which dumps you unceremoniously onto what is inexplicably called the Miracle Mile, an arid, treeless, concrete gateway to a limitless expanse of shopping malls and discount stores that blanket both sides of the road.

I would set out with a list of things—silverware, television set, mixing bowl, a tomato—but I rarely returned home with any of the things I had set out to get, or, if I did, it was only after long detours through the countryside, well beyond Best Buy and Home Depot and Appleby's and McDonald's. I invariably found myself driving through the empty streets of the old towns along the rivers, the towns that had thrived alongside the woolen mills, and died when they did. I would slow down as I passed the handsome old abandoned brick mills, perched above the falls of still-formidable rivers, a few shards of shattered glass winking in their empty windows; past the shuttered storefronts on steep streets of deserted business districts, and the occasional brightly lit, newly renovated café, hoping against all odds for customers. These towns were caught in an ebb tide of history, ghosts of an earlier New England, and they brought the area into focus in a way the highly polished, tourist-thronged streets of Woodstock Village, with its five Paul Revere bells and pre-Colonial houses gleaming as white as newly lightened teeth, never could. Hardscrabble and grim, having long outlived their usefulness, they were more real to me than the fairy-tale town with which I had first become enchanted.

One afternoon I went to look at wood-burning stoves, concerned that the heating bills would bankrupt me before the winter was out. On the way home I got lost somewhere along the border between Vermont and New Hampshire. I was on a winding, narrowing road high above the White River, driving past old weathered frame houses with crumbling porches that sagged under the weight of cast-off sleds and children's bicycles, old tires and plastic bins and broken fans peering out of windows in bedrooms no longer occupied; it was a strange day, a mid-November kind of day that had fallen like a stone into the gleam of late summer. The sky was darkening, it had rained off and on, and there was a touch of spookiness in the air, a brooding New England spookiness, and I thought of Hawthorne, and the witch trials, and the angry old fathers ever on the watch for sin.

I thought of Hawthorne again that day when I finally made

it back up the drive to my house, which stood cheerless and sodden in the rain. The author had begun his writing career in the dreary house in Salem, Massachusetts, in which, following his father's death, he had grown up with his mother, two sisters, and his mother's family. Castle Dismal, he called it, and then and there I rechristened my own house. It was no Fortress of Solitude, and I was no Superman, but it made a very nice Castle Dismal, or as a friend of mine liked to dress it up a bit, Château Lugubre. The name change cheered me up. The way a rain does on a hot day, it cleared the air, acknowledged the unhappiness I had been afraid to admit.

Because the inside of the house depressed me, I decided to concentrate on the outside. Tess Riley had been a passionate gardener, and the first summer after I bought the place, the porch was surrounded by a fanfare of yellow tiger lilies in the front, while on the right side, a bank of pillowy white hydrangeas bloomed. At the entrance to the septic field, a wooden trellis supported the efforts of a young climbing rose, and at the edge of the yard, a determined bank of begonias was doing its best to screen off two gigantic white propane tanks that lay like dead beached whales in full view of the front windows.

The front yard, encircled by the unfinished driveway, was home to a half-dozen young fruit trees, which prompted visions of baskets full of apples and pears and plums. In the shadowy beds near the unfinished stone wall, ferns, bleeding hearts, forget-me-nots, and Queen Anne's lace prospered, while near the front door a young climbing rose had begun an ascent that was clearly meant to one day outline the entire entrance in jubilant pink.

It had been lovely, and I had loved looking at it, the pleasure laced with a somewhat jaundiced regret, the way a roué in a 1930s melodrama savors the loveliness of the young innocent he knows he will ruin. I am no gardener: By the time I moved in full-time, the tiger lilies and the hydrangeas were choked with thorny weeds, and the mysterious vine growing up the trellis had begun to engulf the porch. The fruit trees looked sickly and had yet to produce so much as a blossom, aside from the singular apple that had greeted my arrival. The rhododendron, starved of fertilizer and other attentions, had retreated, and the ugly white propane tanks basked naked in the sun. I bought a beautiful basket of flowers to hang on the porch, but the sun soon burned them to a crisp. The only living thing that had so far survived my ministrations was a small tomato plant in a clay pot near the front door.

I had long ago made my peace with my lack of gardening skills; the lawn, however, was another matter. *Lawn* is perhaps too nice a term for the mixture of crabgrass, dandelions, and milkweed that surrounded the house, but whatever it was, it was growing fast. Worse, the cleared field covering the septic tank was shrinking, as an advance artillery of brush and spindly saplings announced the woods' intention of expanding its territory. This was worrisome, being an entirely too accurate reflection of the chaos that was colonizing my general state of mind. The place was beginning to look as if it belonged to an old lady with eighteen cats.

I had to do something. I consulted with Greg, the hardworking, laconic son of the long-established Fullerton clan, and the man who had looked after the place when I wasn't living here and plowed the road in the winter. Greg said that if I wasn't going to put a proper lawn in—and I emphatically wasn't, once I heard the cost—I might as well just weed-whip the place. I didn't know what a weed whipper was, but I dutifully went to the hardware store just east of the village and asked for one. It turned out to be a cumbersome machine that was expensive and dauntingly complicated-looking. I settled for a slender little gadget—also called a weed whipper—that consisted of a long handle with a set of wicked-looking teeth at the end of it.

The weed whipper was no match for the crabgrass. The wicked-looking teeth promptly bent double. I went back to the hardware store and got a machete. The machete worked, sort of, but it was a backbreaking business, and I didn't make much progress. I went back again—by this time I was getting some openly curious looks—and this time I bought a scythe. I was very excited about the scythe. It had a long, thick oak handle, taller than I was, and an enormous curved blade, and it was oddly beautiful in a Grim Reaper sort of way. Then I googled "How to Scythe" and came up with a YouTube video featuring a hearty, red-faced farmer with a German accent who demonstrated the proper technique. It didn't look too hard. At the end of the video the farmer said to be careful, because if you weren't, ho ho ho, accidents could happen. Then the camera panned to the young boy standing next to him, a big grin pasted on his face despite the fact that one of his trouser legs was empty. I was pretty sure it was a joke.

I liked scything. It was rhythmic and deeply satisfying to chop off the heads of those arrogant weeds, and it made me feel like a character in a Thomas Hardy novel. But scything was also really hard and time-consuming and painful for a woman with negligible upper-body strength. I didn't give up on scything but I went back to the hardware store to buy a hand mower for backup.

I met up with a part-time clerk I'll call Jake, who thankfully hadn't witnessed my other purchases. We bonded over the fact that he and his wife also lived off the grid; in fact, they cooked everything on a woodstove, even baked bread in it. By this time Jake, whose benign blue eyes, grizzled beard, and affable 1960s commune smile masked the instincts of a born pool hall hustler, had an idea of what he was dealing with. After I bought the hand mower, he asked what I was going to do about dead branches—if I had a bow saw, I could use them for kindling. And what about a hatchet, essential for splitting large pieces of firewood into pieces small enough to fit into the woodstove I wanted. I bought the bow saw and the hatchet, and he spent a little time showing me how to use the hatchet. It was all in the swing, he said. Now you try, he said. You should get your husband to do that for you, though, he added. I told him I lived alone and he told me I was very brave.

Pretty soon my garage looked like an armory, as if I were preparing to launch an attack on the place. Or maybe it was a defensive maneuver, a defense of my sanity against that image of the crazy old lady and the cats. I spent hours in the sun, chopping things down, until the lawn was littered with branches and thorny brambles and clumps of severed grass, and anxiety had yielded to aching arms and calloused hands.

I eventually abandoned the long aimless drives, but even the more purposeful ones on more familiar ground emphasized my outsider status. A handmade sign near a lumberyard flashed by, advertising HAY! SECOND CUT! And it bothered me that I didn't know whether the second cut was better or worse, and why it mattered. What was hay exactly? Such details had seemed exciting and curious once, but a visitor sees things differently than a stranger trying to settle down in a place, and Woodstock, as the days began to shorten and the nights to cool, seemed more alien to me than ever. There was so much I didn't know.

One day I had lunch with Lynne at the same coffee shop where I had first felt the urgent need to belong to this place. Over turkey soup she mentioned she would have a great deal more mulch this fall than she could ever use. Would I like to buy some from her?

For some reason that was the moment when the strangeness of what I had done and where I had landed became overwhelming. Mulch? What the hell was mulch? I thought about faking it—I have all the mulch I need right now, thanks . . . I've decided not to mulch this season . . . mulch doesn't go with my furniture—but I didn't have it in me.

I don't know what that is, I said. And stupidly I started to cry. People keep telling me I'm brave, I told her. Why do they say that? Do they know something I don't know?

You're a woman living alone in the middle of nowhere, Lynne said. Brave is polite for crazy.

The more foreign Woodstock became, the more time I spent in the house. I tried to establish some order, but the floor remained a maze of half-emptied boxes. One black high-heeled pump sat on the kitchen counter waiting for its mate to appear. A small sea of crumpled newspaper stirred with the breeze, eddying around stacks of books waiting for a bookcase. Instead of unpacking, I moved furniture around. Obsessively I dragged carpets from one side of the house to the other, and rearranged the lamps, determined to find a combination that looked cozier. I lugged a huge table upstairs to the room that was supposed to be the study if I ever got back to work, and then back down to the room that was going to be the dining room, if I ever had friends to invite to dinner. But none of the rooms ever looked right, no matter where and in what combination I put things. Only the green leather La-Z-Boy remained where it was, too heavy to go anywhere.

Gradually a kind of lethargy set in. The drowsy late-summer heat, the humming of the cicadas, the tumble of creek water over its rocky bed, and the rush of the wind in the trees were a narcotic, a lulling babble in my ears drowning out the imperative to get things done. The emptiness of the house was a presence of its own, an oppressively silent contrast to the murmur of life outside.

I dreaded the evenings. I had craved solitude, but what I had found instead was a loneliness that pressed like a stone on my chest. Sometimes I was almost grateful for that stone; it kept me weighted down when I was sure I would float away, so little connection did I have to the world.

Zoë's departure inaugurated a second grieving for her father. I had prided myself after his death on my independence, declining to confront the awkwardness of middle-aged dating; now I had to blush at my conceit. Without my daughter I was truly alone, and I wondered if one could actually die of loneliness, the pain of it was so physical.

I called my friend Cynthia on the West Coast. We had met in our early twenties, two young writers at the *Washington Post*, and while she was as deeply neurotic about work as I was, she was also the most practical, confident, no-nonsense negotiator of life and all of its pitfalls I had ever known.

I asked her how she had handled the departure of her daughter, her youngest child. I went nuts, she said cheerfully. I knocked on the doors of women I barely knew, women who had gone through this, and said, I don't think I can survive this. I'm pretty sure it's going to kill me. They told me it wouldn't kill me, and I'm going to tell you the same thing, but it will be a while before you believe me.

The tomato plant died.

I tried again to settle in. I swept the mudroom, put some books away, unpacked a box. But the stone would not move, and I was scared. This was the beginning, yes, but the beginning of what? I certainly hadn't seen this coming, this feeling of being punched in the stomach, of wondering whether I could even bear it, whether the grief of Zoë's leaving might be something I could not survive with any degree of contentment. I had never known such nights before, nights that grayed into days that darkened back into nights.

I turned to books for solace. "No man should go through life without once experiencing healthy, even bored solitude in the wilderness, finding himself depending solely on himself and thereby learning his true and hidden strength," Jack Kerouac wrote. But Kerouac died before he hit fifty. What if I didn't have any true or hidden strength? More to the point was poor Alice James, after brother Henry left for England, brother William got married, and her father died. "Those ghastly days, when I was by myself in the little house in Mt. Vernon Street," she wrote in her journal, "how I longed to flee . . . from the

'Alone, Alone!' that echoed through the house, rustled down the stairs, whispered from the walls, and confronted me, like a material presence, as I sat waiting, counting the moments, as they turned themselves from today into tomorrow."

When I was little, the nuns at the Catholic schools I attended loved to tell the story of the ninth-century Irish monks who sailed across the ocean in deerskin curricles in search of a holy solitude, with little more than their prayers to guide them. For me the story had always resonated with possibility, with the faith it took to cleave to yourself, to escape the future as it had been shaped for you in favor of one of your own choosing. But when I looked back, it seemed that I had never done much to influence the path my life had taken. Life had formed itself around whatever canker or happy chance had come my way. I had moved to Vermont to change that, to reach bedrock, the essence of who I was, and to decide for myself what happened next.

But now I remembered that bedrock can be treacherous. Take away the stuff under which it is buried—the topsoil, the dirt and roots, the living things that tunnel beneath the surface—and there is nothing to hold back the drenching rains that can carry away everything of value.

One Sunday afternoon, the soft whisper of the trees outside convinced me it was time to take a walk. Then the quandary: where to go. Not the usual roads, down Route 106 to the country store, or Noah Wood or Long Hill Road. I felt awkward and embarrassed to be out and about on such a fine day by myself, one of the last beautiful days of September, a time when families were gathering outdoors for end-of-season barbecues and picnics. And not the woods behind my house, which were dark and strange—I was so tired of everything that was strange. For days I had listened to the slow drip of rain from the leaves of the trees and the birdcalls and the occasional wild cries and screams. I didn't know the names of the birds, or the trees, or the animals that made those nearly human sounds. Was it better to know the names of things or not to know? When my husband was alive, we had had a summer house in a beach community not far from the city. On Fire Island, I knew, because he had told me, the name of the bird that perched on our roof every day and scolded us when we walked up the path to the house. It was a redwing blackbird, he said, a male, protecting his nest. And because the bird had a name, he acquired in my head a personality and a romance: he was our protector, too, and every spring when we came back he was there as well, our guardian spirit. Then one spring the blackbird didn't come back and that summer Lee was diagnosed, and I hated that bird savagely and grieved for him as well, knowing full well how idiotic it is to find a coherence in two random events, to impose meaning where there is none. We name things so we can know them, and, knowing them, won't be afraid of them. Maybe we should be afraid.

The afternoon wore on. I went upstairs to take a nap, to mute the noisy emptiness of hours that scrabbled like little claws on a dusty floorboard. I lay down on the bed, eyes open, very still, studying a small hole in the window screen that needed patching. And then, something happened. The quality of the air shifted slightly. I was in the same place, in the same moment, listening to the same soft susurrus of the trees in the same warm wind, watching the sunlight slant through the leaves and strike the white paint of the new bed at precisely the same angle it had a moment ago, but something was different.

I heard the heavy thud of a cat jumping down from the dining room table onto a bare wooden floor, the reassuring sound that means you are alone but not alone, that outside the room you're in, the quiet hum of the hive continues. Ah, I thought, Eliot's here, our old fat foundling cat, no doubt walking past the appraising, not altogether approving glance of his leaner littermate, Ezra. And I was happy and relieved because the house was no longer heavy with silence. Then there were other sounds, distant, muted: the screen door from the long-ago beach house banging, the fluting tones of Zoë's Guyanese babysitter, a quick zephyr of giggles wafting up from Zoë's bedroom, which could only mean adolescent coconspirators hatching complicated plots. The contradictory mix of time and place didn't seem strange at all, merely a prelude to what came next, an anachronistic jumble of scenes and visitations, as if the chapters of my life had been tossed into the air and fallen to the floor and been hastily reassembled in no apparent order.

Zoë came into the room as a seven-year-old, all scabs and bones and frayed ankle bracelets, her hair the despair of every comb it ever met. You know, she said, standing by the bed, I had a really happy childhood. But your dad died, I said, and she nodded her head, a little absentmindedly, as if she had told me what she had come to say and was already thinking of what to do next. And then the smell of hot concrete and suntan oil, the tinny strains of a transistor radio, the officers' club swimming pool at the army base where my father had been stationed when I was thirteen. I saw myself walking self-consciously toward the diving board in a ruffled two-piece swimsuit, the one my mother made for me because back then you couldn't buy a two-piece swimsuit small enough for a skinny thirteen-year-old with a flat chest. And the girl I had been rolled her eyes at me, as if to say, I know, I know, it's silly to feel so pleased with myself, but she *was* pleased and a little proud of how she looked at that moment, at the very beginning of a brand-new world. And then the scene changed again, to a large cocktail party in Washington, D.C., and it took me a moment to find myself, but there I was, in a large group of friends and rivals and colleagues, laughing, flirting, reaching for the glass of Jack Daniel's I would regret the next morning. You see, said Zoë, who had wandered back, you were liked, and I said yes, a little surprised. And then she said: You were loved. I looked closer and saw the faces of old friends who were dear to me still and then looked even closer and there was my husband and I said yes, humbled now to realize I had been so lucky so far beyond my merits. And then I was in a car in the darkness, driving to a town I had never seen and would never see again, to write a story, deeply thrilled with the adventure of it, a young reporter on her own. And then in a room I didn't recognize, blanketed by the night, making love with someone I wanted very much, and I felt the desire of it as the drug it used to be back then, as if diving off a high, high cliff into the blue of a bottomless sea. And then I was sitting on the sheet-covered sofa that was the best sofa in my parents' house, all dressed up, waiting for the end of the last long hour before we went to greet my father, back from his tour of duty in Vietnam. And then suddenly, an altogether different sofa, and Lee was beside me. We haven't sat like this for a long, long time, I said. And then I paused and added, as gently as possible, You're dead, you know, wishing there was a kinder way to say it. And then Lee held me. I felt his arms around me, really felt them, not the way you do in a dream, but the way you do in life, and I was sure it was him, that he was really there. Thank you, I said, and I said it over and over again because he had come back to see me and then I was crying, and the tears were forcing my eyes open. I have to open my eyes now, I told him, and he smiled, a little sadly. I know, he said, but when you do I'll have to leave. I tried hard but my eyes did open, or at least I thought they did, and I was standing on a ledge, rocky and barren, high above the sea, and I was thirsty and couldn't drink and then I opened my eyes for real, and wept for all that had been and was no longer.

I walked downstairs unsteadily, half looking for the images that had been so real, wondering for a second where Eliot the cat had gotten to, before remembering. *I just had the dream you have before you die,* I thought to myself. And for a long while after, I felt like something of a ghost haunting my own life. It would be a longer time still before I saw the thing for what it was—a reminder of the fullness and variety of all that had come before, a valediction, forbidding mourning.

Middle age resonates with so much loss, profound and superficial: expectations die, friendships fade, hairlines recede, looks change, and health and hope are no longer givens. It becomes easy to forget the fullness that has come before; self-pity, while a dreary threadbare flannel when worn by others, has a luxuriant silky feel when we wrap it around ourselves. It was some time before I came to marvel at the images that had presented themselves so vividly, and to be deeply grateful for them. I might have no idea of what would happen next, but it seemed ungrateful to complain when there had been so much that was good in the past, whether or not I had had the wit to recognize it.

I knew then that I would have to find a way forward, despite the fear and paralysis. We create boundaries and then we defy them, like the monks who set sail for the end of the world.

When my daughter was a freshman in high school, her English teacher, a tall, brilliant, sardonic man, would threaten the class on a regular basis with the fate he said awaited them if they failed to turn in their assignments on time: a transcript pockmarked with mediocre grades and tepid recommendations, which would in turn ensure their eventual matriculation to the University of Guam. In such an unabashedly competitive school as hers, and placed as she and the rest of her class were on the woozy heights of the bell curve shaped by the bumper crop of babies produced the year she was born, this was a dire fate indeed.

I never found out whether there really was such a place, or if it existed merely for the purposes of Mr. Bender's ironic humor, but the University of Guam was always lurking at the back of my mind, a dark miasma of doom. I pictured it as a windy conglomeration of Quonset huts and concrete, a place without hope or mirth or future, and to dispel the image, I would turn up the wattage on my nagging, chivvying her about homework or urging her to join résumé-friendly clubs and debating societies in which she showed absolutely no native interest or ability. None of this had any effect on Zoë, who had spent her elementary school years at a small "progressive" school in Greenwich Village that did not believe in tests or report cards or, for that matter, arithmetic, and she still subscribed to the astounding notion that grades didn't matter and learning did. It was an admirable principle that rendered the usual process of surviving high school and applying to colleges as crazy-making as possible, but in the end it worked out just fine.

Because, as it turned out, Zoë didn't go to the University of Guam.

But I did.

3

Henry

In those days, I lived lightly in the half-empty house, walking softly on the bare pine floor, so as not to hear the echo of my own footsteps, making as little noise as possible, like a polite houseguest searching the kitchen for a train schedule while the hosts are still asleep. It was not my house, but it was not the Rileys' house, either; it seemed to belong, if it belonged to anyone, to the woods, and to its emissaries, the spider who lived in the mudroom, the mouse I found in the cupboard one day, to the silence I was reluctant to disturb. The long golden days stretched themselves like cats, then stalked off into the evening without a backward glance.

The only way to make the house mine, I knew, was to take possession, to unpack the boxes and put things away, to claim the space for my own. And to that end, I would wake up feeling strong and centered; I would open a box and get as far as spreading its contents out on the floor, the better to group them according to some sort of purpose, something I had failed to do when packing up. Accordingly, laddered black tights, printer cartridges in yellow and cyan, a whisk broom, silk scarves, a half-dozen books, a sweater I hadn't worn in years, and a cascade of Post-it notes, paper clips, hair combs, and several cartons of Mickey Mouse Band-Aids all made it into the same receptacle.

The disorder was daunting; trying to ignore a rising sense of alarm, I would scoop up an assortment of more or less office-related objects and head into the upstairs bedroom that was to be my work space. It was a small room that overlooked the meadow, perfectly adequate as a place to write, but utterly insufficient for the titanic battle that raged within it.

I had always been a messy person, and I had come very late, to the extent I had come at all, to the idea that there might be something wrong with living in a state of unending domestic chaos. I had grown up in the late 1950s and early 1960s, the decade social anthropologists now define as the zenith of the American preoccupation with dirt and idealization of the perfect homemaker. Middle-class suburban homes bristled with veritable armories of cleaning devices while television ads promised spotless perfection through the aid of muscular genies emerging from toilets (a terrifying prospect to a young child) and magical bubbles that danced away the dirt. And these homes were inhabited by a generation of smart, ambitious, energetic women condemned to staking their personal worth on the shininess of their sinks and the spotlessness of their kitchen floors.

The combination was a volatile one, at least in my house, where every Saturday was witness to a clash of biblical proportions between my brash, infuriated, and infuriating mother, and a horde of witless, unbelieving Philistines in the persons of my sweet-natured but essentially indolent father, my two brothers, and myself.

To my mother, the weekly Saturday-morning cleaning was a holy crusade. To the rest of us, it was the tar pit that separated my father from the golf course and his children from the morning cartoons. It always ended badly; by the end of the day my mother was a tornado of recrimination, exhaustion, and injured feelings, and she was nearly, but unfortunately not completely, speechless with anger. Housekeeping and rage: back then they went together like Sears & Roebuck, gin and tonic, death and desire. My mother scrubbed, and cursed our laziness. She was ferocious in her quest for perfection. Any spot, any lint left behind was a slap in the face. On weekdays, the enemy was the dirt itself. But on Saturdays, the one day when she assumed she would have allies in the cause, it was our inexplicable indifference to her vision. She was smart and possessed of boundless energy and will. She should have been directing the fate of nations. Instead, she had us.

By the time I went off to college I loathed housekeeping and gloried in the disorder of my dorm room as proof that my roommate and I would not suffer the same fates as our mothers. Housekeeping was not only politically incorrect; it could literally drive you crazy, if the film *Diary of a Mad Housewife* was any indication. (I still remember the obnoxious social-climbing husband sneering to his daughters, "Your mother made Phi Beta Kappa at Smith, but I don't think she can make a four-minute egg.") Besides, we had better things to do: stop the war, save the world, talk about boys. So I reveled in our squalor, reading Charles Dickens and rolling my eyes at his plucky little heroines whose principal virtues inevitably revolved around their tidiness and their abilities to turn even a prison cell into a domestic paradise, while listening to Mick Jagger bragging about his unmade bed. Mick Jagger or Little Dorrit? The choice wasn't even close.

Marriage to a man with an innate love of an organized life had changed all that—at least cosmetically. Our clashing attitudes could have led to an instant divorce, but housekeeping was never really an issue—Lee was a man with a quiet but indomitable force of will. I think he simply intimidated our home into staying orderly.

After he died, however, the old ways returned, and in time the apartment in New York was in shambles. I had meant to make a fresh start in Vermont, but the simple process of unpacking a box was enough to make my heart race with anxiety. Every item that I unearthed—an old grocery list in Lee's hand, a pair of socks I'd worn when I went into labor, a crayoned drawing of Simba from *The Lion King*—became a totem, at once sacred and mocking. That's when I would break down, filled with grief, yes, but also fury. I could see that the mess wasn't simply the product of my lifelong laziness or some outdated adolescent protest. No—I was drowning in the past, the artifacts of loss fossilizing around me. By never letting go of anything, no matter how trivial, I was dismissing the future, allowing it no room in which to unfold. No wonder cleaning up caused me such anxiety.

And yet I couldn't do it. The house developed claws and fangs and ripped away any notion I had of gaining purchase, and I would slide fast into the old grief, past any sense of purpose. Defeated, I would creep downstairs, curl up on the sofa, and listen to Emmylou Harris, too tired to do anything, too empty to feel guilty about it. "Hold on," sang Emmylou, but I couldn't, not very well. I tried to think of a reason to live, not in a melodramatic way—as George Eliot said, suicide is attractive only to the young, and besides, there was Zoë to consider, Zoë who had recovered so well from her father's death, but who just the week before had called the Woodstock sheriff and convinced him to drive out to Castle Dismal to make sure I was still alive. (The phone was dead, and she hadn't been able to reach me.) No, this was a question asked of the future. If Zoë was all but gone, and writing was too frightening, and love a distant memory, what was I going to do with myself? How did one invest this part of life with meaning? Now what? I didn't know, and Emmylou didn't say.

Every night I promised myself I would climb out of this trough and get back to work, and late every afternoon I would discover to my great surprise that I had done nothing at all. Instead I knitted—I was making a red throw for Zoë's room in college—and while I knitted, I thought about what I would write when the throw was finished. On the computer, I played endless games of Boggle and Word Whomp, a word-scrambling game in which animated gophers poked their heads out of holes. When you failed to find the big seven-letter word in the allotted three minutes, they shook their heads in disappointment, but when you got it right, they would do back-flips and ecstatically munch the turnip awarded as your prize. I loved making them happy.

The word games were the closest I got to writing. Some of the words I unscrambled would hook me with their inherent beauty or the chime of some subconscious resonance. Sometimes I would write them down—*gaunt, goblin, bramble, languor, dwindle*. I played for hours, until my fingers hurt. Then I would go on the Internet and research addiction to computer games, looking for symptoms of how badly I was hooked, until at last the rooms began to darken and the day was down for the count.

I don't know how long this state of suspended animation would have lasted if the ceiling in the mudroom hadn't caved in. One morning I woke to find fluffy balls of acid pink insulation drifting through Castle Dismal. I wasn't worried about them so much as I was alarmed at the idea of armies of mice pouring through the breached ceiling to complete their final conquest of the house. I called Larry Davis, the contractor who had done some earlier work on the place, and we agreed that it might be time to look toward some of the other repairs that had long been in the offing—a roof over the garage to keep the snow off, an enclosure for the solar battery, a new shed for the woodpile, a set of stairs down from the door that currently led nowhere, a rerouting of the electrical wiring that was housed in a giant column in the middle of the kitchen, making it impossible to see from one end to another. A crew would arrive on Monday.

I told myself that it was a terrible imposition—the noise and the dust and the bulky presence of strangers would keep me from writing, and I did a pretty good job of pretending to resent that fact. But of course it wasn't true. I was sick with lack of work, with not even knowing where my winter clothes were. I had two days to make room for the men and their equipment. The mudroom had to be cleared of its maze of boxes, and by the time they arrived, I had finally made a dent in setting up a kitchen, and filling a bedroom closet, and clearing a tentative space for books and papers.

There were four of them: Mike, young and handsome, a

high school football star who had turned to drugs after an injury sidelined his dreams, only to sober up at eighteen when his girlfriend gave birth to their son. Calvin, his stepfather, small and strong and wiry, in his sixties, pared down by the race life had run him, but observant, weathered, a stealth wisdom playing in his eyes. Hank, gap-toothed, bashful, taking refuge in hammer and nail and the sense they made out of everything. Jimmy, round-faced, potbellied, a little lazy, hiding it, or so he thought, behind an easygoing temperament.

We were elaborately polite to one another for a day or two, making wide berths in the course of our passing, until one afternoon I sneezed—I make china rattle when I sneeze—and Mike, who was on the roof, sang out, "God bless you!" and I shouted a thank-you and we all laughed. Then it was easy and the house rang with the noise they made, and vibrated with their energy and the authority they brought to bear on the obdurate house, pounding, sawing, and thwacking it into submission, chasing away the wretched silence. They worked hard, those men, and their energy shamed me into trying a little harder.

While they worked, I would settle down to read, but instead I listened to the talk about lives so different from mine. They talked about women—Mike was moving into the trailer with his stepfather and his mom; he and his girlfriend weren't getting along. He worked days, and she worked nights so that one of them would be with their son, but they never saw each other. And the coming hunting season—the winter had been hard and the deer scarce and the freezer still needed to be filled if they were going to make ends meet. How stupid my inertia would seem to them, if I were to join the conversation: *"Well, my daughter left home, you see, and I don't know who I am now, and the days fit like a coat three sizes too big . . ."* It would sound like Urdu.

A few weeks before the men had arrived, I had brought home the puppy I had contracted for in the spring, in New York. He was a fat ball of fluff, with a big head and no shape to speak of, and his hair was so white and thick he looked more like a baby polar bear than anything else. A dog had always been an essential element in the full Fortress of Solitude fantasy. I had planned to name him Carlo after Emily Dickinson's dog. Like Dickinson (before she became a recluse), I would wander the country with only my dog for company, and I would write great things that no one would see, and be at one with nature.

But if I was no Emily, the dopey little pile of sleep I had brought home from New Hampshire was no Carlo. He had, to the extent that he had any expression at all, a mild, unintelligent look, as if he might grow up to be a docile country curate, the poor, unambitious kind who yearns after the pretty girl in Victorian novels. Henry, I decided. He was a Henry.

So far our life together had been rocky. I'm not sure what had possessed me to get a dog—I had once had a wheaten terrier, and it had been a nightmare. That decision had been equally well thought out—Zoë, aged six, had lost a beloved stuffed dog, and coming so soon after her father had died, the loss loomed large; it seemed imperative that I go out and immediately buy her a real one as a replacement. The result was a turbocharged, highly aggressive puppy in the hands of an incompetent and unmotivated owner who thought the creature's crazed disposition was kind of cute until she wreaked absolute havoc in the apartment, as well as on my friendships.

Two threatened lawsuits and three trainers later, I was told by the last one in no uncertain terms that the dog was too nuts to live in a city, and so Rosie went off to live on a beautiful farm in Vermont, where she has lived happily to this day.

Henry would be different.

Henry *was* different: Henry hated me. It had been fine the first few days, when he would fall asleep next to his food bowl and I would carry him to the sofa and hold him while he napped. But then he woke up.

Henry didn't want to play with me. He didn't want to cuddle or for that matter be touched. He didn't like any of the pile of educational toys I bought him. He ignored all my attempts to house-train him. We would walk around the front yard for hours—at dawn, after every meal, in the middle of the night—while I encouraged, pleaded, threatened, and begged him to do what he needed to do, without result. Then he would toddle back inside and shit on the floor. When he wasn't comatose or incontinent, he was half shark, shredding magazines and furniture and shoes and electric cords. Most of the time, I didn't even have the will to stop him, so completely oblivious was he to my presence. Except once.

I remember that day because it was one of those moments of rare insight when you realize just how crazy you have become. It was late on a hot afternoon, another day of getting nothing done. The sun was just beginning to disappear over the ridge, for which I was grateful—it was the hour when the fury at myself for doing nothing and the anxiety that prevented me from doing anything about it finally dissolved into a promise that tomorrow would be different. It was always a tricky transition, however, and sometimes the lie didn't take—I knew that tomorrow wasn't going to be any different, and I would spend the evening writing long journal entries about what a failure I was while listening to old Cowboy Junkies albums.

But that evening, I was simply relieved to have the day done, and I went outside to sit on the stoop and play with the sad-faced puppy. He was worrying a bone I had given him earlier and ignoring me and my efforts to interest him in a game we could play together. So I decided to take the bone away—a move, had I known anything at all about dogs, I would have executed with extreme caution (if at all), but I didn't know anything about dogs. Besides, the dog-training book said it was important to establish dominion over your puppy or he would always be the master, and thinking of the lamentable Rosie, and also perhaps of my inability to establish any dominion over myself, I was determined to do just that.

I crouched down next to the oblivious ball of fluff. Drop it, I said, in the low tone of command I was instructed to use in puppy-training class. Henry didn't look up. I said it again and again, and of course he ignored me. Finally, determined to exercise at least this much authority, I grabbed the bone to take it away. Henry snarled as savagely as something that looked like a stuffed baby polar bear could snarl, and bit me.

For a moment I just stared at the puppy, who looked back at me, his head slightly cocked, endearingly, infuriatingly indifferent. I had been trying so hard to make him love me. Then I sat down in the dirt and cried. That got Henry's attention, and he stood up. But if I was expecting a Disney-esque moment where he loped over to console me—and of course I was—I was mistaken. Henry edged away, with a little frightened whimper, turned his back on me, and fell asleep.

As I sat in the dirt, half laughing at the utter misery of the moment, a memory bloomed, of rolling around on the floor of the loft with Zoë after a checkup at the pediatrician's. I hadn't slept in days, Lee was in China, and I could not figure out how to get her out of the Snugli that bound the baby to my chest. She was crying and I was crying and I was pretty sure that in the history of the world in all its folly there had never been so spectacular a failure of a mother. Zoë's arrival had changed everything, and I had wondered if I would ever be the same. I never was of course, a fact for which I am still stupendously grateful.

Now, Zoë, by her leaving, had once again turned my life upside down. But if her presence had become, after our somewhat shaky start-up, the wind in my sails, her departure had blown away my moorings as easily as a child blows away the seeds on a dandelion, scattering everything I thought I knew, leaving an emptiness unlike any I had known. Why was I so undone?

Because she was the only thing I ever did right.

The idea came unbidden. I ran my hand through the dirt, feeling its grit on my palm, breathing in the dank air that smelled of the decayed fallen weeds, looking up at the opaque, tree-shrouded hills that seemed every day to grow closer and closer, making it harder to see the sky. Was it true?

Raising Zoë had been the only thing I had done without design. As with Henry, I had read a million books about how to be a mother, but Zoë had thrown them all out the window simply by virtue of being herself. No book could tell me who she was, no parenting guide could tell me what she needed. Only she could do that.

Being a mother was the only thing I had done instinctively, improvising, making mistakes, figuring out what worked and what didn't. I had never set out to be a great mother or a good mother—I hadn't even known I wanted children until I fell in love with my husband's. I expected nothing from myself—I just didn't want to mess her up. The British pediatrician Donald Winnicott had a phrase for that: he called it being a good enough mother. And I think I had been that.

What had saved me—what had saved Zoë—was that motherhood was a world I had entered without expectations—my own or any others. If I had tried to be a great parent I would have failed, or if I hadn't failed I would have figured out a way to belittle my performance, to blame myself for botching the job. But unlike work, or school, or any other arena of life that mattered to me, parenting wasn't an arena where goals had been set—by myself or my parents or anyone else—and not met, where the need to be good had inevitably made me run headlong the other way. Zoë was chaos, the wilderness, where I had had to make my own way.

That was the way the past looked to me then, staring past the snuffling puppy in the fading sunlight, to the maze of memory that was my map to the way things were. For a long time it was the only map I had, but it was not in fact the only one that existed.

The next morning Calvin and Mike and the others had arrived to fix my roof, the mudroom, and the woodshed, and, as it turned out, to restore a little sanity into the bargain.

For one thing, they loved Henry. They didn't mind him getting in their way when they worked, they brought their wives and girlfriends to meet him, roughhoused with him when they had the time, and ignored him when they were busy. They laughed when he rolled in mud puddles and cuffed him when he misbehaved. He was just a dog to them, not a litmus test, not a sociopathic shark puppy, just a dog doing doggy things. Calvin, in particular, took an interest in him. He had grown up on a farm and helped his father raise forty hunting dogs. He's a fine pup, Calvin said. He's going to be a good dog.

I worry that he doesn't like me, I confessed.

Calvin gave me the look, the slightly pitying but polite you're-talking-to-a-foreigner look I was getting used to in Vermont. He has to, Calvin said. Beat him till he's blind, he'll still think you're the sun and the moon. Can't help himself.

One afternoon, I came outside to find them staring up the driveway at a corner of the yard where the wild weeds met the more feral brambles that marked the onset of the creepy woods.

They were watching Henry, they told me. Every afternoon he ambled up the circular path of flattened grass that was the putative driveway, headed for the brambles, stayed for a while, and then half trotted, half tumbled back down again. Something up there sure got his attention, they said.

I walked up after him. The puppy would stick his head in the bushes and jerk it out again, nibbling something. I looked closer. Blackberries. The brambles were actually blackberry canes, and that corner of the yard was thick with them. But getting at the fruit was hard work for the puppy: most of the berries were too high up for him to access and the ones closer to the ground were deep within a wall of very sharp thorns.

I reached in, snagged a few of the fruit, and tried them. They were fat, sweet, and juicy, blackberries as I had never known them. I grabbed a few more and knelt down and offered them to Henry. He snaffled them up and we looked at each other, in mutual greed, and I went to work, dividing the impromptu harvest equally until we had both had our fill.

After that, our trip to the blackberry bushes became a ritual, first thing in the morning, last thing in the afternoon. The puppy began to follow me, and together we explored more of the land immediately surrounding the house than I had ever done on my own. On such frail anchors do we begin to hang a routine, a rhythm, a connection, perhaps even a life.

One day I came back from some errands to find Calvin, Mike, Hank, and Jimmy packing up their gear for the last time. The work was done: the ceiling in the mudroom was freshly painted and perfect, the wood was back in the newly constructed woodshed. The redesigned garage now boasted a space big enough for a car and a roof that would withstand the snow. There was nothing left to do (actually there was a lot left to do, but there was no money left with which to do it).

I was sorry to see them go. Sorry, and a little afraid of facing the empty house and its silent rebuke. As he was gathering the last of the paintbrushes and plaster knives, I thanked Calvin, the de facto leader of the group, for the work they had done and for helping me to understand Henry. He's a good dog, he said again. I think he's ready for more. Take him out, maybe in the woods. I had the feeling that perhaps it wasn't just Henry who Calvin thought might be ready for more.

It was turning into a spectacular fall, one of those rare intersections of changing weather and waning light combining to produce the most vibrant color seen in years. In town, the residents and the shopkeepers were as stunned by the display as the leaf peepers, as they called the tourists, and drivers would stop their cars in the middle of the roads leading in and out of town, gobsmacked by the beauty.

All that loveliness—inviting, ephemeral—made the woods much less forbidding. I stood at the back window, watching the play of light on the leaves of a white birch. Calvin was right: Henry and I were ready for a walk. The puppy had never seen the creek; we would clamber down the hill at the back of the house and take a look around.

I found Henry under the crab apple tree. Ever since our blackberry communion, he had been more tractable. A day or two earlier he had ambled into the living room with something loathsome in his mouth. Drop it, I had said, more out of habit than of hope, and much to our mutual surprise, he did.

I took Henry to the backyard, to the lip of the hill that sloped steeply down to the little creek. At first I was wary: I had rarely walked in these woods without getting lost. But the falling of the first of the leaves had begun to create gaps in the formerly solid wall of green, piercing its opaque mystery. If we didn't go too far, if we kept the house in sight, surely nothing could happen.

At first we stuck to the plan. The puppy ran down the hill until gravity and speed overthrew him and he somersaulted and slid on his belly, bouncing off stubbly bushes and outcroppings until he flopped into the creek and immediately set about chasing dragonflies and darting little fish. Then a shadow on the other bank caught his attention and he was off, splashing through the shallow water and up onto the far side of the creek.

I started across the creek myself, intent at first on simply fetching him back. But it was a lovely afternoon, and the warm wind seemed charged with a volatile excitement, and throughout the wood there was a sense of kinetic energy, like that of a great orchestra tuning up, playing on the infinite variety of shape and color, light and motion of the trees. I was suddenly aware of being at the beating heart of a moment, of the leaves high above my head that were just barely rimmed with crimson, their stems still stained with the green of summer, and the ones just a breath away from falling, curled and brown, their history written. It was the kind of day that mocked your fears, made you impatient with any and all hesitation. On such a day you can do anything; I was suddenly sick of the anxiety and the lethargy that had lingered for so long. Today was the file hidden in the layer cake, the key slipped inside a book beneath the wary eye of the guards.

Henry was almost out of sight. I took a last long look over my shoulder, and up the hill, back the way we had come—I could still see the house, sort of, through the trees. We would be fine, I thought, mostly because I was far too giddy to think otherwise at that point, and because it was too fine a day for anything bad to happen. And besides, there was something else: I knew things now that I hadn't before. Or thought I did.

I had brought two books with me to Vermont. One was a standard navigational handbook on the uses of map, compass, and altimeter. I had cracked that one first. It started out encouragingly enough: any moron, it promised, could learn to find his way in the woods. That was the shallow end. The deep end of the book was made up of passages that talked about how to combine map reading with compass use: "After the compass has been set for the proper bearing, carry the compass and the map in the same hand, holding them firmly together with the edge of the compass lined up with the route of the map and the direction of travel arrow pointing forward. Then keep the compass map and (one)self oriented by always keeping the north part of the compass needle over the north arrow of the compass housing, letting the front right hand or left hand corner of the base plate take the place of the thumb." *Take the place of the thumb?*

The second book was called *Finding Your Way Without Map or Compass*, written in 1957 by Harold Gatty. Gatty, whom Charles Lindbergh called "the prince of navigators," was a handsome, daring, and distinguished Australian flier who in 1931, with his partner Wiley Post, circumnavigated the world in a record-breaking eight days, a feat so stunning that the United States awarded him the Distinguished Flying Cross by a special act of Congress.

After World War II, in which he had served as group commander in the Royal Australian Air Force, Gatty and his wife settled in Fiji, where he established an airline and set up a coconut plantation on Katafanga, his own island. At the same time, he traveled the globe, researching his book on path finding as it was practiced in every culture and climate, illustrating his observations with detailed drawings of camel caravans in the Arabian desert, termite mounds in northern Australia, Maori canoe routes, and glacier tables at the North Pole.

Gatty had trained bomber pilots for the United States Air Force, and the responsibility of sending young men out over trackless oceans, deserts, and snow-covered tundra lends his book an underlying sense of urgency and precision. At the same time, his work is a prose poem to the glories of the natural world and the mysteries it veils in plain sight.

Gatty recounts, for instance, how the explorer and traveler F. Spencer Chapman was once kayaking with an Inuit hunting party along the east coast of Greenland when they were enveloped in a thick fog. They were close to shore and could hear the sound of the waves breaking, but Chapman worried about how they were to find the narrow entrance to the fjord that would take them back home. The Inuits were unconcerned; "indeed they beguiled the time by singing verse after verse of their traditional songs," he wrote, "and occasionally they threw their harpoons from sheer joie de vivre."

After hours of steady paddling, the leader of the group suddenly swung the kayak toward shore and hit the narrow inlet right on target. Chapman was amazed; he spent days trying to figure out how the men had done it. It was only on subsequent trips that he finally twigged to the secret: the Inuit had been listening to the songs of the male snow buntings that nested along that coast. Each bird sang a slightly different song, and the Inuit recognized the notes of the birds that nested on the headland of their home fjord.

Sir Ernest Shackleton and other polar explorers made their way to camp by steering their sledges at a constant angle to the sastrugi, the parallel ridges of snow that invariably ran north to south, much like the sand dunes by which the Bedouin navigated in the Sahara. Pioneers on the American prairie oriented themselves by way of the pilot weed, whose leaves grow in a north-south direction. Longfellow refers to the same plant as "the compass flower" in *Evangeline,* his epic poem about the expulsion of the Acadians by the British. It was the plant, he wrote, "that the finger of God has suspended / Here on its fragile stalk" to guide his heroine across the emptiness of America as she searched for her lost love.

In southern Europe, you can orient yourself, or could, in Gatty's time, by the vineyards, which were planted on the southern sides of the hills in order to take full advantage of the sun. In less cultivated country, the traveler could look to the mountain ridges, which are less eroded on the northern side. On the banks of the lower Orange River in the mountains of Namaqualand, hundreds of plants incline their heads to the north. In Afrikaans, they are known as "half mens" or "half men" because from a distance the plants look like silhouettes of people buried up to their waists in the ground. In Fiji, the flower plumes of sea reeds grow on the side opposite to the direction of the prevailing wind.

Clouds over land can point the way to water; clouds over water can point the way to land. Look up: the dark shadow you see on the underside of a cloud formation may be a water sky, the reflection of a pond or a patch of snow, while a brightness of light on a dark cloud over the ocean in northern waters might be an ice blink, the reflection perhaps of an unseen iceberg ahead.

If you are lost near Tristan da Cunha, the Falkland Islands, or southern Argentina, the presence of large numbers of great skua flying overhead can tell you how close to shore you are. The Polynesians, on long night voyages over hundreds of miles, frequently carried pigs with them—the pigs would get excited when they smelled land. *Shabur,* a shimmering haze in the desert that points the way toward an oasis, is visible ten feet from the ground, the same height as a traveler's eye when he is sitting on top of a camel.

I started out reading Gatty's book because I thought it might offer a shortcut—who needed to know how to read a compass if the world was such an open map? But soon I was reading the book in the late afternoon, curled up in the chair in my bedroom, making my way through it as if it were a book of fairy tales, skipping the detailed charts and the careful caveats, fascinated by the pageantry of strange facts and remote places, and the keen observations of this intrepid man and the other adventurers.

What I learned, sitting in the overstuffed rocking chair, was that the world was large. Looking back, it seems to me now that in reading that book, in thinking about the messages to be found in weed and rock and a prevailing wind, a large door began to open, a chink of light let in.

Not all of Gatty's observations were esoteric; I learned, for instance, that the pattern of growth on a tree can tell you which way south is, and that such a pattern held true as much for the sunstruck pines of Virginia as for the windblown cypresses of southern France. Gatty accompanied this wonderful fact with a drawing of an elm tree in Holland. The right side of the tree faced southeast and was full of branches, while the other side was undernourished and skimpy by comparison. How obvious! I began to swell with confidence.

Now, it might have helped, given the course of subsequent events, if I had read the paragraph directly under the drawing of the asymmetrical elm, which clearly stated that "when observing a tree for directional purposes, it is essential to choose a tree growing in an exposed position and one which is not interfered with or sheltered by other trees or buildings." But I missed that part.

Instead I checked the sky and the position of the sun. It seemed to me that if I simply kept the sun over my left shoulder, I would have no problems. A basic knowledge of astronomy might have been useful here; I might then have considered the fact that knowing where south was didn't much matter if you hadn't stopped to figure out where your own house lay in relation to the sun, or for that matter, that if you keep the sun on your left for any length of time, and the sun moves westward, as it is wont do, then you are going to be traveling in an arc, not a straight line. Instead I kept Gatty's drawing of the elm tree firmly in mind: there was nothing ambiguous about it—one side was full and geometrically perfect, the other practically anorectic in its foliage, like some of the young women in Washington Square Park who would shave one side of their heads and leave the hair on the other side long and wild and iridescently purple.

We wandered up the hill, more or less in a straight line, but I wasn't paying too much attention. The falling leaves, crimson, yellow, still vibrant with life, the snuffling, scuffling puppy rooting in the bushes, a shaft of sunlight illuminating an unsuspected stand of white and yellow birches, the musky smell of fallen tree trunks and mushrooms and moldering damp, the sharp scent of the pine, the bright bite of a wildflower—there was no time to pay more than glancing attention to the sun or any other marker. I had fallen into the timelessness that the woods evoke, existing as they do in a different tense altogether, one that is entirely present and yet comprises both its ancient and its most recent past: the glittering of mica in a rock that might have arrived on the tide of a glacier, a fallen feather still smelling of the cloud that passed along shaft and vane.

We don't wander much these days—in fact we are rather afraid of it, rarely venturing across town without a GPS or a smartphone that can pinpoint our precise location. We are a long, long way from the footsteps of the philosophers, of Rousseau, who walked across France and Switzerland to escape his critics and attackers. "These hours of solitude and meditation are the only ones in the day during which I am fully myself and for myself, without diversion, without obstacle, and during which I can truly claim to be what nature willed," he wrote in *Reveries of the Solitary Walker*. Thoreau, that old curmudgeon, was quite uncompromising on the correct and incorrect way of walking in the woods. "We should go forth on the shortest walk, perchance, in the spirit of undying adventure, never to return—prepared to send back our embalmed hearts only as relics to our desolate kingdoms."

The seventeenth-century Japanese poet Bashō, a virtuoso of the art of wandering, saw the unplanned path as a way toward wisdom, insight, and beauty. He was part of an already ancient tradition: to wander was a Taoist metaphor for ecstasy. Bashō couldn't stay put: even though his disciples built him several huts and planted a banana tree in his yard in the hopes of inducing him to stay, he spent much of his life on the road. He traveled on foot, and took the longer, more difficult, more dangerous route. He always left home, he said, expecting to be killed by bandits or to die in the middle of nowhere. His journeys took him over hundreds of miles, for months at a time; he nearly died of cold in the mountains while riding down rain-swollen rivers. It didn't matter: wandering was the ideal way to find and to appreciate the transcendent ephemeral moment, to let it offer up its beauty.

The soft panting of the puppy caught my attention; he was lying spread-eagled in a pile of leaves, too tired to even look up. The sun was blazing hot high overhead, and I was thirsty, very thirsty. How long had we been walking?

I looked around. We were standing at the bottom of a hill, two hills, really. I looked up in the direction from which I thought we had come, but I couldn't see the house. Of course—we had crossed the stream and started up the other side. And then? Where had we gone then? Scrambled images came to mind—a bridle trail, a stone wall, a scrap of pink boundary tape wrapped around a maple—but they bore no relation to one another, told me nothing about the order in which I had seen them.

I raced up the higher of the two hills, ignoring the scratches of brambles and raspberry canes, crashing through briars with Henry in my arms. From the top, I would figure out which way to go, I would catch a glimpse of a road or a house, perhaps even my house. But there was nothing to see at the top, only more trees and another ridge.

It went on that way for hours. I crossed and recrossed the brook, followed deer trails and the occasional bridle path and crumbling stone walls that abruptly ended. The sun was so hot that I was afraid the trusting little creature by my side would expire before I got us back home. I tried to take my cues from the trees, as suggested by the dauntless Gatty, but they had nothing to tell me—in the darkness of the woods, there was no clear-cut message to the foliage patterns, or at least none that

I could discern. Moss, which I'd read somewhere always grew on the north side of a stone, displayed a remarkable flexibility on the issue in this part of the world. The sun—well, I couldn't remember where the sun was supposed to be or on what side of it I was meant to be, and anyway, how could the sun be in the south when it was directly overhead? How was that even possible?

I fought down a shiver of fear. After all, there were hours left until dark. But the country was remote; these woods, endless uninterrupted acres of them, belonged to people who rarely came near them. No one would hear me if I shouted. No one would even know that I was missing.

We were lost and we stayed lost for hours. I yelled for help, a little hesitantly, because I wasn't sure which was worse, getting no response and wandering for hours until exhaustion set in, or getting a reply, and having to face life as the Woman Who Got Lost in Her Own Backyard. Fear triumphed and I yelled louder, but I heard nothing, not even a distant generator or the drone of a tractor. I sat down with the puppy in my lap, too tired to move, hot, hungry, afraid.

When I was young I loved to wander, though I never pretended that my motives were dressed in any of Bashō's transcendent purpose. To leave was the thing, and never to arrive. Wandering was an escape, from every inconvenient obligation and follow-through, every unpaid bill and neglected friend and sulky lover, from the thoroughly annoying inconvenience of myself. My twenties were a topsy-turvy decade of confusion, exaltation, and despair, one in which I had no idea what I wanted, or how to get it if I did. Sometimes I gloried in my

reckless and improvidently romantic approach to life; more often, I worried that I was lacking in some essential element of
character, that somehow I wasn't as real as other people my
age, who had begun to grow up, who made plans, honored
commitments, and seemed to possess a clear sense of where
they were going.

I was living in Washington, D.C., at the time, working on a
daily newspaper, wanting passionately to do well despite a harrowing conviction that I was a talentless fraud, and that one
day my editors would see the mistake they had made in hiring
me and cast me out of Eden. It made for a rather contingent
kind of life, in which dread was kept at bay by large quantities
of Jack Daniel's and various less-than-legal chemical aids.

When it all got too overwhelming, I would get in my car
and pick a way out of town and drive and drive, the windows
open, the radio loud, the direction a matter of supreme indifference, driving until the miles expunged every anxious, guilty,
self-loathing thought. As the day waned and the high-beamed
lamps of the highway switched on and the dusk began to ob-
scure the shoulders of the interstate, I would look for the most
likely exit, ignoring signs for the familiar motel chains huddled
close by the major exits in favor of the more obscure-looking
off ramps leading to a small strip of rooms with inadequate
lighting and old beige or brown wall-to-wall carpeting, with
just enough room for a bed and a TV set placed on a bureau,
a patchy neon sign outside indicating you had arrived at the
Wayside, or the Cozy, or the Starlight.

Looking back, those trips don't seem as purposeless as they
did then: they were in some respects my own version of Proust's
madeleine. My family had taken a number of cross-country

trips as we moved from one military assignment to another, and it was during those long hours in the family station wagon, the fresh promise of the motel at the end of the hot and dusty days, that I felt the safest, the most at home.

But there was more to it than that: setting out for nowhere inevitably involved getting lost at some point, and I liked that part. Sometimes the confusion of time and place led to serendipity and to adventures in which the ordinary transformed itself into extraordinary beauty. Then the country in which I wandered became a dreamscape in which visions, marvelous and troubling, could be conjured from the commonplace.

One morning, after a sleepless and misspent night, I got in my car and headed south toward Chesapeake Bay, because I needed to see the ocean, because at the moment when the sun had made the lamplight redundant in a room in which I had never meant to stay until morning, the ocean seemed like the only thing that could rescue me from the tidal pull of my own foolishness.

Several hours later I at last wandered into a twisty, shadowy road that led deep into scrubby pines. I thought the road would take me to the water's edge, but it dead-ended at a swamp instead. The car sunk up to its hubcaps in the mud. Wandering on foot, looking for a road that would lead to rescue, I came across a small, sunlit, grassy meadow strewn with red wildflowers, in which stood a very pregnant chestnut mare. On an ordinary day, in the ordinary way, it would have merited a mere flick of the eyelids, but I was in the rapturous state that frayed nerves, heightened senses, and a jolt of adrenaline can create, and the beauty of the scene, the animal's glossy flanks and quiet calm, and the absolute silence rioted through my young and addled head, dissolving the fear and insecurity on which I had been choking. In that moment, I knew that everything would be all right; I could look past my own doubt and see that the world was immense and my future as rich in possibility as the swelling belly of the brown mare. I remembered that moment always. It was a great gift, a promise that was redeemed, many times over.

As far as the real world went, I was still lost, of course, and would be for hours afterward, but that was mere detail. When you are young, time is buoyant; you know that sooner or later it will float you home.

I thought about that girl, the one I had been, and how nothing about the situation I was now in would have bothered her. She would not have felt this vulnerability to chance, to the insouciant indifference of fate. But then nothing at all had happened to her yet. So this is what it means to get old, I thought, to find yourself at the mercy of your own mortality.

One of my projects in coming to Vermont had been to learn how to grow old, not merely gracefully, but also with style and panache—and I had convinced myself that I could take it all in stride, with a minimum of lamentation. But until that moment when I found myself lost, in my own backyard, I had never really understood what we lose when we finally realize we are not immortal. It didn't seem that far a leap to the fear that flashed in my grandmother's eyes whenever she crossed a threshold—because, she said, you never knew whether you would make it back.

The shadows were lengthening. I began to wonder if we would have to stay the night. The very thought inspired new energy—I walked back up a hill I'd tried before, determined this time not to stop until I had reached some ridge or crest from which I might gain some perspective. Finally I came to another bridle path, which was well worn and led downhill— surely it would lead eventually to some human habitation.

Eventually the path intersected with an unfamiliar stream, and I changed course once again, deciding that the stream was the better bet—if I followed it down, I told myself, I would eventually come to a house. I did, about twenty yards later—my own. The house was not where it was supposed to be. It was nowhere near where it was supposed to be. Later, much later, when I was looking at a topographical map of the area, I could get a general idea of what I had done, which involved circumscribing a giant loop de loop of about eight miles. But back then all I knew was that I was tired, and deeply, deeply angry.

I wasn't angry because I had been lost. I was angry because I no longer possessed the optimism and the elasticity that had made getting lost a delight, and if I could no longer lose myself, this troublesome burdensome self, in the way I most loved, then I did not know what to do.

I sat on the steps of the porch for a long time, the exhausted puppy in my lap—Henry had been so tired that he had fallen asleep with his head in his water bowl, sending forth a small flotilla of bubbles with each exhale. It was strange: I had not minded growing older until now. None of the obvious drawbacks of age—the lost looks, the creaking knees, the constricting horizon—had seemed particularly upsetting, compared with the relief to be had in surviving one's youth.

But the disastrous walk reminded me that I had reached an age where only careful planning and a steady eye would keep me safe—in the woods, and in my life. I needed to stay

alert, to find my bearings, if I was to avoid wasting time walking in circles. I couldn't stumble into old age the way I had
through my front door, not quite knowing how I got there.
And yet, in a way that I could not define, the best of me—as
well as the worst—was inextricably tied up with a love of the
accidental and unexpected. I had never been the sort of person who made five-year plans, or saved for a mortgage, or
even kept a date book. I admired people like that, I envied
them, but I had never wanted to be like them. I had led a
life in which I had made few thoughtful decisions, and yes,
it had cost me dearly in many ways, but it had also brought
me great happiness, and in the end it was simply who I was.
For me, the miraculous had forever been bound up in the
random; what if the elements crucial to growing old in a way
that wasn't self-deluded were antithetical to what I needed to
be happy?

I was not the first to freight a sense of direction with so much
moral weight, to bind the knowledge inherent in compass and
map to the dictates of philosophy. Direction has always been
bound up with magic and with faith.

The first maps pointed the way to heaven; the compass began in China as a fortune-telling device. In the thirteenth century, an English sea captain confided to his journal during a
long voyage that he had to take great care when using his compass aboard ship. So great was the power associated with this
new device that were his sailors to see him use it, they would
fear him as a magician who used black arts to set their course.

Even those who did not fear the compass saw it as much
more than a navigational aid. Jacques de Vitry, the Crusader

bishop of Acre, acknowledged the device's usefulness while sailing almost as an afterthought—he seemed far more impressed by its resistance to witchcraft and poisons, and its use as a cure for insomnia.

The poets, too, were dazzled. Erasmus wrote that geography was essential to poetry, and it's easy to see the compass as a kind of poem, marrying as it does the molten core of the earth to the steadfast position of a distant star. In the thirteenth century an Italian poet, Francesco da Barberino, wrote a poem advising his readers on how to lead a good life in the aftermath of a shipwrecked heart. Build a compass, he advises. It will point you to virtue. In the *Inferno,* Dante contrasted the dangerous lure of physical love with the soul's inner compass, which points the soul to God's eternal love. Ever since men first tried to get from one place to another, it seems there has been a distinctly moral quality to knowing where you are going, to choosing the right direction.

The compass, of course, makes no distinctions among directions. On its plain round face, the choices are weighted equally; east is as good as west, north no better than south.

And yet, from the beginning, the four cardinal directions have been symbols of good or evil, weighted differently in every time and place and culture. To say that a man had gone west, in an English novel of a certain era, was to say that he had died—Golgotha, where Christ was crucified, lay to the west. Resurrection, on the other hand, lay to the east, the Orient, which is why, when we get lost, we must first orient ourselves. Indeed, before the Reformation, many European maps placed east, not north, at the top, while in China, south was at the top, because south was the location of the gates of paradise.

In Bali, however, believers faced north to pray, and the word *insanity* in Balinese translates as "not knowing where north is."

Every direction is freighted by a shifting amalgam of history, culture, and religion, a tug of the senses at a subliminal level. North is forbidding, harsh, purifying, dreadful. South is indolent, seductive, ripe. East is mysterious, of course, and mystical. West is shining with possibility in this country, full of woe in others. These connotations, these dreams of past and future, influenced the placement of temples, the resting place of a suicide, the augury to be found in the flight of a bird.

Which way to go? What choice to make? Nowhere is the profound anxiety of direction more evident than at a crossroads. No other geographical feature bears the same freight of human mystery, fear, and speculation. The intersection of one road with another in every culture portends a meeting with doom or destiny. A place to meet the devil or see the man you will marry, to watch the witches dance or confound the ghost who is out to get you.

There is something poignant about this need to invest direction with such significance, to find benevolence emanating from one path and evil from another. We have always needed to know not only where food is to be found, and fire and shelter, but also which way leads to heaven and which to hell. In direction, symmetry, in the elegant phrase of the geographer Yi-Fu Tuan, is pulled apart by life—because life itself has only one direction, and we fight like hell not to go there.

I wasn't looking for the gates of paradise, but I had a lot riding on my own understanding of direction; in my own way, I hoped, like the poet, that it would lead me to virtue, and like the sailors, I was more than a little afraid.

Waypoints

Adventure is a sign of incompetence.

—VILHJALMUR STEFANSSON, *Arctic explorer*

Al-Hakim bi-Amr Allah, third Fatimid caliph and sixteenth Ismaili imam, after riding into the Musett hills on his donkey for one of his regular nocturnal meditations.

The Ninth Spanish Legion, during the Roman conquest of Britain.

Vandino and Ugolino Vivaldi, Genoese sailors, while attempting to reach Asia from Europe by sea.

Gaspar Corte-Real, looking for the Northwest Passage.

Francisco de Orellana, while exploring the Amazon.

Henry Hudson, set adrift by mutinous sailors.

American politician John Lansing, while mailing a letter.

Thomas Lynch, after signing the Declaration of Independence.

Ludwig Leichhardt, on the Darling Downs between the Swan and the Conmore rivers, in Australia.

Henry Every, pirate, after capturing the flagship of the emperor of the Mughal empire and escaping with £600,000.

Percy Fawcett, in the jungle of Mato Grosso, searching for the lost city of Z.

Judge Crater, after entering a New York City taxicab.

Everett Ruess, while sketching in the Utah desert.

Paula Jean Weldon, on the Long Trail, near Glastenbury Mountain.

D. B. Cooper, from the rear of a Boeing 727 with $200,000 in cash.

The three lighthouse keepers on the Flannan Isles.

Dorothy Arnold, heiress, after buying a book in New York City.

Arthur Cravan, French Dadaist, near Salina Cruz, Mexico.

Richard Halliburton, while sailing a Chinese junk across the Pacific Ocean.

Michael Rockefeller, in southwestern New Guinea.

The crew of the *Mary Celeste,* their ship discovered, seaworthy and under full sail, near the Strait of Gibraltar.

The lost haunt us. The famous, the notorious, the hapless, the determined, the ones who walked out of their lives and never looked back, the ones whose fevered dreams consumed their money and their health and their lives, the ones who acted on a whim that might have been forgotten a moment later had a door blown shut or the sound of a dinner bell distracted them. The ones who got away with it. The ones who didn't. They remain immortal, living on in mystery long after they would have survived in life. They conjure complicated emotions— not merely grief, but envy, anger, frustration, bewilderment. Such is our need to know.

The lost haunt us: our helplessness, our loneliness connect us to theirs. There are so many fates into which our imaginations cannot enter, but this one binds us; we all know what it is like to be lost. The fear that comes with being unmoored from the familiar, the free fall of the spider from the web of identity, the recognition of just how precarious one's hold on life is. You can feel it on a mountain trail, but you can also feel it in a parking lot in broad daylight. To be lost is to not know not only where you are but also who you are, to realize how much of your sense of self is embedded in place, in context. We see ourselves by the light of the familiar. Take away the familiar, loved or loathed, and confidence evaporates.

So perhaps it is not surprising, then, that the stages of grief and the stages of being lost are much the same: the lost and the grieving stumble along the same double helix of what they knew and what they must come to know, trying to relocate themselves along the same fragile axis of understanding.

In the beginning there is always denial. You refuse to believe that you don't know where you are, that you can't find your way, that anything has changed. You keep going the way you had determined to go at the outset, walking faster, perhaps, and with more urgency, certain that if you simply press on, then everything will fall into place.

After denial, anger: the grieving blame the dead, or life, or God. The lost blame themselves, or the compass, or the woods. The anger drives out logic. The anger leads inexorably to the third stage, to the frantic search for a way to make the present emergency conform to the old reality. In grief, it's called bargaining—if only you will bring him back, if only you will let me live, I will do anything. In the woods, it is the time when the lost waste what energy they have in frantic attempts to assure themselves that the lake they can see from the top of a tree is surely the lake they passed an hour ago. The fork in the road is the same fork they were looking for all along. This is the time for walking in circles; this is the time of dread. The grieving person is trying to find a way back to the life he knows. The lost person is trying to make the map in his head, his mental map, fit the world he is looking at. But it will not. It cannot.

Depression is next. You are beaten. Everything you know, about yourself and the world and the way things could or should have been, has no bearing here. You give up, at least for a time. It's not a bad thing. You are doing what you have to do—you are letting go of the world as you thought you knew it, in order to see the one that confronts you now.

After depression, acceptance. A tricky phase, because it can lead you to darkness or to light. Either you will surrender to your grief, and let time stop and twilight take you, or you will bury your dead and take the first tottering step into the unknown. Either you will curl up in a tree trunk or an abandoned school bus and wait, or you will decide that it is up to you to find yourself, because only after you have done that do you have a prayer of finding your way.

When I first got back to Castle Dismal after that disastrous walk, I was sure I would never again travel down any surface that didn't have a dotted white line down the middle. But before the alarm and confusion had given way to Overwrought Existential Despair, there had been that small intense euphoria, found in a bright button of a wildflower, or the cool spookiness of a hollow tree, a sense of elation tied to nothing but the moment, and of that, I wanted more.

There had to be a way to walk away from the house when it was driving me nuts, away from the questions and the turmoil and the paralysis in a way that didn't court utter disorientation.

There was one possibility. At the bottom of the hill, my

road, Keeling, as it was then called, intersected with Noah
Wood, which, to the left, led out to Route 106, the two-lane
blacktop that ran north to Woodstock and south to the interstate. But to the right, Noah Wood ascended sharply uphill before essentially dead-ending. Past that point it turned into little
more than a trail much like my own. But I had no idea where
it went.

I thought of asking Greg Fullerton, but I hesitated. Greg,
though polite and patient, had an air of such competent self-
containment that I was always a little embarrassed to admit
the full extent of my haplessness to him, though it was no
doubt on full display: he was, after all, the person to whom I
turned for the kind of questions that weren't even questions in
his book, and most of our early encounters involved him pulling my car out of the ditch whenever the Jeep took a flyer. My
neighbor Tom was also a possibility, but we weren't on the best
of terms at the moment.

Tom was a native of the area, a boy from one of the neighboring villages who had moved away to Massachusetts and
made good with his own investment company. He had built a
beautiful house up on a hill on the ten-acre parcel next to mine
and used the place mostly on weekends, in summer when the
haunting sounds of his alto sax floated out over the hollow between us and pretty much nonstop during deer hunting season
in the fall, when he had been known to take potshots at bears
from his deck.

Everyone who knew Tom said he was the nicest, most ten-
derhearted guy on God's green earth, which made it all the
worse that I seemed to invariably drive him into a state of
tight-lipped indignation. When I used to visit my house inter-

mittently, he had been a kind and friendly neighbor, but lately everything I did seemed to drive him nuts. I was a flatlander, that was part of it, I knew, and I was pretty sure he thought I was nuts living in so isolated a place by myself, an opinion that made me furious, only because I was in imminent danger of sharing it. But there was so much I couldn't know and didn't see at the time about this complicated, sensitive man with his own inevitable set of misgivings, sorrows, regrets, and doubts. He came up to this part of the world for escape and for solace, and an inept newcomer bumbling around wasn't high on his list of the place's natural wonders. Which meant that most of our encounters had involved lectures, half serious, half teasing, about the general wrongness of everything from the house itself to my way of living in it. It was a stupid place to build a house! It was in the middle of a swamp! Did I know that? The driveway was in the wrong place and at the wrong pitch, my dog pooped in the wrong place, that wasn't the way to haul wood into the house, no one could figure out what I was doing up there.

Recently we had had a major set-to. Tom had discovered Henry with an apple in his mouth. The apple was one of dozens that had fallen from a tree at the side of our mutual road, but the tree was definitely on Tom's land.

Tom was in a bit of a bad mood—he'd been busy blowing up beaver dams in reprisal for the beavers' felling of a two-hundred-year-old Vermont apple tree; he kept repeating the description, as if two-hundred-year-old-Vermont-apple-tree was a genus of its own. So the sight of Henry munching happily on one of the windfall apples that had tumbled to the ground from one of his still-standing trees was the last straw.

Halfway through a lecture about how the apples were for the deer because the deer ate the apples and hunting season would be here soon, I had interrupted him. I know all about that, I said eagerly, needing to show him I wasn't totally uninformed about the customs of the country. I had heard about it from one of the men working on the house. He was a bow hunter, I told Tom; in fact, I had invited him to hunt on my land.

Hunters, it turns out, are somewhat territorial about their hunting grounds. Tom's eyes threatened to leave their sockets. He told me I was never to do that again and conjured up images of throngs of armed men running amok, gunning down what were apparently his very own personal deer. I wanted to point out that state law permitted hunters to hunt anywhere they pleased on land that wasn't posted to the contrary, that we lived next to a state forest in any event, and that it wasn't really up to him to tell me who could or could not hunt on land to which I held the title. But I was too taken aback. I asked him, incredulously, if he was angry. I gotta go, he said, and zoomed off, muttering something about how people like me were ruining Vermont, how pretty soon the place would be so left it would be illegal to make a right turn.

I'm a coward when it comes to confrontation, and besides, I didn't need anyone else telling me what an incompetent I was: I was doing just fine in that department on my own. So I had been avoiding Tom as much as possible. But it was the middle of the week, hunting season hadn't yet begun, and he probably was in Massachusetts. I decided to chance walking down the road, take the mysterious right onto Noah Wood, and see, finally, where it went.

It was a misty morning of soft light and little movement in the trees; only the rushing of the little brook that paralleled the road broke the silence. Along the side of the road the last of the summer wildflowers bloomed, scraps of lavender and orange against gray stone wall and black lichen. There had been a bit of rain the night before, and the sudden dips were slick with wet and the air was sharp with the scent of cool air rushing up under the warmth. I began to enjoy myself, hopscotching down the rough parts of the road, using my walking sticks to swing myself over the rocks.

At the bottom of the hill Henry and I turned right, onto Noah Wood, which would climb steadily upward for three-quarters of a mile before petering out into a dirt road. I didn't know how far the Jeep trail went, or where for that matter, but at least it was a marked path to follow, which meant that this walk was unlikely to end in savage self-recrimination. I could always turn back.

There were two houses at the intersection of my road and Noah Wood, both inhabited only occasionally by their out-of-town owners. But farther up, after a steady climb to the crest of the hill, there was a tidy, deceptively compact two-story frame home built snugly into the side of the ridge, which then ran down to a large pond. To the left was a small brown barn encircled by a split rail fence, along which the last of the summer's lilies shot out a few forlorn salutes. A chestnut horse and a pony of the same color looked up at us with mild curiosity. Henry was thrilled and slipped under the railing to make friends, at which point the side door of the house opened and a small, fast-moving blur burst onto the deck, shouting warnings about the horses and the dog and the dire results of a collision between them.

But the horse and the pony decided to tolerate Henry, and their owner invited me in for a cup of tea. Her name was Harriet Goodwin. She was a clinical psychologist and gerontologist from Boston who, with her husband, Dean, a mechanical engineer, had retired to South Woodstock over twenty years ago. She was a petite woman with piercing blue eyes that radiated shrewd assessment and a dry humor and arch manner that made her somehow appear taller than she really was. We had met briefly on one of my earlier visits. It was impossible not to get to know Harriet; she was frank, curious, talkative, and opinionated. Most of the residents of Woodstock, particularly those who had moved here from somewhere else, were polite but reticent to the point of blandness, at least until they knew you better, but Harriet was much too interested in people to observe such niceties. She knew everything about the area and everybody in it, and every newcomer was fair game for a thorough cross-examination, after which she was able to pinpoint your precise location within the South Woodstock social geography.

This time around, Harriet hauled out Dean for a proper introduction. He was a giant of a man, as quiet and grave as Harriet was talkative, with a handsome, fleshy face, his head slightly cocked as he listened intently. Dean was one of those people, and there are few of them, with whom you know right away that only your best and most plainspoken self will do. He wasn't much for small talk: he would listen, to see if there was any help he could provide, and if there was not, he would vanish and leave the ladies to their conversation.

I told Harriet about Noah Wood Road and asked about the possibilities of getting lost there. Hah, she said, with her short bark of a laugh, people go up and never come out, swallowed up by the confusion of the bridle paths and snowmobile trails and dried-up streambeds that snaked their way around and through the densely forested ridges and the watersheds between them. But Harriet had ridden those woods for years, and knew them well. Slowly, and with many repetitions at my request, she plotted out a simple and, she promised, plainly visible route that should take me in a circular fashion back to my house. It looked simple enough, but Harriet wasn't taking any chances. Call me when you get back, she said. Just so I know you made it.

The path she described took me along the broad rutted back of a dirt road, which if I missed the turnoff, would eventually take me over the hills to the north and deposit me on the other side of Woodstock. We didn't. After about a mile or two we came to an intersection with an unnamed but clearly visible path that cut through boggy marshland before opening up onto a slightly more open landscape of overgrown meadows, broken stone walls, and an occasional apple tree—all that was left of someone's efforts to make a life there. Off to the sides, deer trails led into chapel-like glades; their calm beauty was inviting, but I was too nervous to explore.

It was a longish walk, about two hours, ample time to worry whether I might have misunderstood Harriet and taken a wrong turn. But she had not overestimated my abilities, and eventually we passed a large, high-banked oval pond that was familiar. It belonged to Therese Fullerton, my nearest full-time neighbor. Therese was a quiet, elegant, silver-haired woman, a recent widow, and mother of the reliable Greg. She and I lived nominally on the same street, except for the fact that a stretch of state forest grew between us, with only a footpath providing any continuity to the two parts of the road—a fact I had not registered until I went looking for my house for the first time after I had closed on it. I was standing in the middle of what turned out to be her driveway when she drove by and pointed me in the right direction.

Just after Therese's house, the trail emerged from the woods, checked in briefly with the paved roadway of Long Hill Road before heading back into the state forest and then descending in bumps and jerks until it rounded a slight curve and there, there! was Castle Dismal, gleaming in the sun. I was relieved to see it, glad to be home, too relaxed to worry for once about the walls closing in.

"There is an unaccountable solace that fierce landscapes offer to the soul. They heal, as well as mirror the broken places we find within." Belden Lane, a Presbyterian minister, was writing of desert and mountains, wild places, where transcendence is etched in emptiness and harsh extremity. There was nothing fierce about the wood that Henry and I took to walking in after that first successful outing; it was a homely place, a scruffy collection of commonplace pines and maples and birches, punctuated with the poignant reminders of modest dreams long forgotten—a mossy cellar hole, an old road that once led to a schoolhouse, interrupted by the flash of a flame-red maple leaf pasted by the wind to a weathered stone. Except during hunting season, I never saw another person while I was walking through them, but still, these were tamed woods, owned and used by others, posted with NO TRESPASSING signs and cables strung over private roads, a battered outhouse, even a sign promising pizza up ahead at a popular snowmobile crossing.

But the solace was there. Henry and I walked along that same path nearly every day, no matter the weather, and the woods became to me an intimate and yet deeply mysterious place, as changeable and strange and strangely familiar as the course of my own thoughts. We wandered past the fallen trunks of massive trees, stripped of all their leaves, wispy strands of fog caught in their bare blackened branches. Sometimes they looked like fallen warriors and sometimes like prehistoric spiny-backed creatures, and sometimes like galleons resting at the bottom of the sea. The living trees changed as well, light and weather and mood painting them in varying shades of hope and dread. They looked like pawns, or gibbets, like soldiers standing with fixed bayonets, like the twisted creatures of nightmare. Sometimes they bent in arabesque like dancers, sometimes their branches traced the outlines of anguish, and in the wind their leaves trembled, fluttered, fought, exulted, surrendered. I was at home there. I didn't know why and I didn't ask. But there was something powerful there in that shadowy wilderness, and in the quiet watchful witness of its unseen and unknown inhabitants, whose presence was betrayed only by a sudden skittering through the underbrush, or the scolding chittering from a branch too high to see its outraged occupant. In the woods, I, too, was hidden from everything that was scary or confusing or hurtful; in the woods, loneliness was replaced by solitude, and self-doubt gave way to curiosity. I kept to the trails, always, but for the most part that was a comfort, not a constraint. I was free, even from my fears. In the woods there was no one to be, no future to plan. And there were moments of such distracting beauty that my own worries subsided to their proper level of inconsequence.

One morning as I started down my own road toward Noah Wood, I ran into Chip Kendall at the bottom of the hill. I didn't know Chip well, but I liked him: we had first met when he came to the house late one night to let me know that there were search parties out looking for a couple of men lost in the woods, so I shouldn't be worried if I heard unfamiliar noises. That's too bad, I'd said. Were they hiking? The back woods were notorious for their twists and turns—once a horse had gotten lost in the hills and was found a week later on the other side of the ridge fifteen miles away. No, said Chip. Hikers, I can understand. These guys were in a rig. They were—he paused, and a look of infinite disdain had crossed his face— GPSers: a couple of guys in possession of a souped-up rig, a gadget with an automated voice telling them to go down a non-existent road, and the common sense God gave a moth.

Everyone in the neighborhood had at least one such story, which were always told with a mixture of cheerful contempt and native pride that this was not an area that would ever take the technological bit between its teeth. Harriet had told me a story about a woman who had nearly gone off a cliff while towing a horse van because the GPS told her to take a dirt road into the woods at ten o'clock at night. The trailer was halfway off the ledge before she had realized her mistake.

Chip owned the sugar bush—an eighty-acre stand of maple trees—across the road from Castle Dismal. A little ways down the hill, the wood was threaded with plastic blue tubing that provided a conduit for the sap during the spring sugaring season, which then collected in the huge metal drum at the side of the road. He was from an old Woodstock family and served as head of the volunteer fire department, a man in his early fifties, with a direct gaze, a thatch of black hair and a black mustache, and a manner so laconic he made even the most inveterate Yankee seem loquacious. But he had as well a quiet air of authority that didn't make itself known until called up by circumstance—when the floods came in the wake of Irene a few years later, Chip was everywhere at once, marshaling the crews, mediating with the town, getting the bulldozers where they were needed most. He drove my neighbor Tom a little nuts because he routinely refused to help with the road repair on our fragile bit of thoroughfare. It was Tom's contention that Chip's big sugaring trucks chewed up the bottom of the hill. Chip countered that his trucks had been able to get to where they needed to go long before anyone decided to live there. Chip's argument cost me money, but I liked his unapologetic logic: he took the woods on their own terms; even a road was a bit of yuppie decadence as far as he could see.

Chip and I exchanged hellos that morning as I walked by. You off to see your friend Harriet? he asked. I was. I didn't ask how he knew—up here everyone always seemed to know everything. There's a shortcut, he said, explaining that if I hiked up the ridge through his sugaring trees, I would come out in a meadow on the other side of Noah Wood from Harriet's house. I tried hard to understand his directions—walk into the woods from this landing, head south and take that deer trail, angle a bit west—but my brain froze at the absurdity of ever remembering all this stuff, or recognizing a deer trail if by any stroke of luck I happened to find one. I confessed to being directionally challenged, so he drew me a crude map on the back of an envelope, which I put into a pocket and went on my way. Thanks, I said; I'll have to try that.

Chip went back to work and I continued down the road. I couldn't tell him, because to a man who had lived here all his life it was laughable, but his assumption that I was capable of negotiating the woods with no trails or blaze marks or any indication of what direction to take was like suggesting I fly to Harriet's on my own two wings.

Still Chip's idea lingered, a small brightly colored lure that took up residence in my imagination and made its presence known every time I walked down the road. What if I could do as he suggested? What if I could simply head off into the woods as casually as if I were walking down a city block, and get to where I wanted to go? Or set out to go anywhere in particular and still make it back? What would I have to know, who would I have to be, to manage it? Small, easily ignored but stubbornly present, possibility bloomed by the side of the road.

One morning I woke up to a wild roaring. A cold wind was charging through the trees like a pride of lions while ragged clouds raced along overhead and the torn remnants of a fog clung to the lower branches like some half-remembered dream. Henry, wet with rain from a brief dash outside, burrowed under the blanket beneath which I huddled on the sofa until the damp weight of wet dog forced my attention to a practical concern: I was freezing. I dumped the puppy on the floor and set about making a fire in the new woodstove.

In the end what broke the paralysis of those first few months was as simple as that. The days were growing colder and I needed to stay warm, a need that demanded no explanation or justification beyond creature comfort. The woodstove I had bought to replace the efforts of the small and drafty fireplace had been delivered weeks before. It was a Vermont Castings stove, to my mind one of the single most elegant devices ever made, and not just because the company gave their models names like Defiant, Intrepid, and Resolute. It was a rare combination of beauty and ruthless efficiency, if you knew how to work it—I had seen similar models in action in houses all my life, though I had never been the keeper of the flame.

I went looking for the user's manual, which I knew to be somewhere near the stove itself, but which was now buried beneath a tottering pile of books. Finding it wasn't all that helpful: I couldn't even manage to make it through the glossary of terms. I cruised past air-to-fuel ratio before giving up somewhere in the middle of back puffing, baffles, and burn rates.

But the cold is a wonderful teacher. I got a sickly fire going that first morning, but over the days and weeks that followed, the fire got stronger as I became more adept. Gradually the fire became the spine around which the body of a day could organize itself—from the necessity of getting it going first thing in the morning without even the comfort of a cup of tea, to the bundles of kindling to be gathered during the afternoons, to the logs I hauled in from the woodpile at the end of the evening to refresh the supply. I began to learn the stove's rhythms and quirks, how and when to feed it in order to keep the temperature steady and the blaze in check.

A friend of mine once remarked that to live successfully in the country in Vermont involves a conscious step backward, if not into the nineteenth century, then at least to the first half of the twentieth. There are a lot of things that don't work, or work slowly, or don't exist once you are outside the confines of the towns and villages, and you can drive yourself crazy if you don't come to terms with that simple fact. I had no cell service; my Internet, until DSL arrived, was at the mercy of a voracious porcupine and worked infrequently, and the telephone functioned mostly as stage dressing—the nearest telephone pole was a half mile away, and the telephone company, for a bewildering number of reasons involving bureaucratic regulations, state and local requirements, and the rights of individual property owners, had declined to provide service. What communication I did have was due to the fragile length of telephone cable Mr. Riley had laid along the side of the road, where it was subject to backhoes, large trucks, Weedwackers, and small things with sharp teeth. The post office didn't deliver, and the FedEx guy tended to leave packages in a ditch at the bottom of the road. A lot of packages didn't make it at all—for a long time, when visitors entered my address into a search engine, the Internet volunteered the information that no such place as my address existed. As an occasional visitor, I found all of these things charming. As a permanent resident, they made me nuts.

Learning the ways of the little Defiant, sitting like a squat toad in the trim brick fireplace, began to change all that. I began to move with the rhythms of this more demanding, more physical way of living, where light and weather dictated the day's routine. In the mornings, on days that promised sunshine, I would do the laundry. The afternoons yielded earlier and earlier to shadow, and the low light that shot under the clouds struck everything gold before swallowing the house in darkness, save for the one pool of light where I worked or read.

The coming cold meant other challenges: I was not the only one who spent more time indoors. In the evenings, I began to hear a low thunder overhead in the rafters, like a tiny herd of migrating wildebeest. I thought I could live with the invisible hordes, as long as they stayed invisible, but one evening I opened the kitchen cupboard to find one of my coinhabitants looking back. A mouse stood on the shelf, holding a bit of cracker in his paws. He peered at me curiously, as if to ask if there was something he could do for me. I was flummoxed. I thought of mice as timid creatures, instantly put to flight by a human's terrifying immensity. But the mouse didn't move, not even when I yelped and jumped back about a foot and a half. After that we stared at each other for a beat. You might at least have the decency to act scared, I said out loud.

I was willing to ignore one mouse, but of course there wasn't just one mouse, and they became increasingly bold—one ambled across the kitchen floor, right past the nose of the sleeping Henry, who opened a single eye to follow the little beastie's progress before going back to sleep.

I got lots of advice, all of it pretty drastic: kill them, kill them before they take over the world! At the hardware store there was an entire shelf devoted to mouse eradication, catering to every temperament. There were poisons that induced varying degrees of painful deaths and traps of all sorts, some of them covered in glue, which left the victim alive to starve to death, some of them spring-loaded, which afforded a quicker end. Some exposed the corpses to view; still others functioned more like little hot-sheet motels, where the mouse crawled in and you didn't have to see what happened next, though you could hear the struggle. There were electronic devices that promised to drive the mice away with a high-pitched sound, and organic sachets full of herbs that mice allegedly hated. My friend Lisa suggested what she called "The Stairway to Heaven": she and her boyfriend had devised a little gangplank that led up to the lip of a toilet in their country house. The mice crawled up and drowned in the toilet bowl, where they could easily be flushed away.

I bought all of them. I tried the electronic gadget, which simply wasted electricity I didn't have, and the herbal sachets, which the mice treated as palate cleansers. But the poisons and the traps all stayed on the shelf.

It was hard to dislike these mice. They didn't look or act like New York mice, all scruffy and worn and harried from the stress of making it in the big city. They looked like Disney mice, small and plump and glossy and bright-eyed. They were absurdly cute; I couldn't bear to hurt them. Besides, they were company. I wasn't proud of this admission, and I was pretty sure it qualified me as the ultimate pathetic flatlander, until I met a man at a dinner party much later. He was a tough, native-born guy, a hunter, a Vietnam vet with a reputation for no nonsense. But when I met him he had softened considerably—he had recently married again, and he still looked at his bride with the expression of someone who simply couldn't believe his good luck. Before he met her, he said, he had been so lonely that he had made friends with one of the mice in his house. Every evening he would put a piece of cheese on his coffee table so he had some company while he read the paper.

Dean, Harriet's husband, suggested I try the West Lebanon Feed and Supply Store if I was looking for the kind of traps that left their occupants alive. He mentioned the place as if it were a store like any other store, but in fact West Lebanon Feed and Supply, to a newcomer, was an Ali Baba's Cave of Wonders, exotic and marvelous, one that catered to every rank and degree and nuance of the animal and vegetable kingdoms, addressing every peril and perk of life in the country.

You could buy anything at West Lebanon Feed—horse vitamins and dog food and boots for mucking out stables, leashes and bridles and fireproof gloves, ointments that would heal the crumpled horns of breeding cows, or repair the inflamed anuses of baby pigs, and complicated devices, festooned with pulleys and cranks, for pulling balky lambs out of their mothers. There were flowering bulbs and fertilizers and potions to encourage the most finicky plants. There were bird feeders of every variety and different kinds of seeds designed to attract songbirds and hummingbirds and nesting birds. Simply walking down the aisles made me happy. I left the place with a half-dozen Havahart live-capture mousetraps, more toys than Henry would ever need, and a fifty-pound bag of birdseed.

I laid the traps in strategically chosen corners of the kitchen and the pantry and I caught a lot of mice in the beginning. But mouse trapping is a labor-intensive business—mice, I learned, must eat frequently, so if you didn't want them to die of starvation, the traps had to be checked frequently and the mice released quickly. On the other hand, you couldn't let them go in the front yard, or they simply made a beeline for the house. For a while I got up every morning and walked down the hill with my traps to release the mice into the woods, where it probably took them all of an hour or two to make their way back. Then, suddenly, the mice disappeared. I rejoiced—could they possibly have learned their lesson and moved on? No, as it turned out. The next time I went out to the garage I found the contents of the fifty-pound bag of birdseed spilling onto the cement floor, courtesy of a neatly chewed hole at the bottom of the sack. I went to fetch the broom and dustpan but stopped short. It occurred to me that the mice and I had found a compromise: the mice stayed out of the kitchen, and I stayed out of the garage.

That first fall in Vermont was like learning to live all over again, figuring out what it meant to take care of myself. It's a lesson we sometimes need to learn more than once. I had a friend whose girlfriend left him after a relationship of many years. He was in shock; he hadn't seen it coming. He had met her in college, and they had lived together ever since. Now he was almost forty. I have to figure everything out, he had said, shaking his head at the enormity of it. This morning I got up and went to make coffee and then I wondered—Am I the kind of guy who makes espresso? Or am I the kind of guy who buys his coffee from a truck on the street? He had no idea who he was on his own.

I had learned how to stay warm, and that initial attempt at comfort led to others, equally basic, equally important. When I had first moved into Castle Dismal, I stopped making real meals for myself, partly because it had been hot and partly because it didn't seem to make sense without Zoë. I lived on cans of tuna and frozen dinners. But now I began to cook again, comfort food for cold weather: squash soup and baked apples, lentil stew, chicken with rice. And for me, cooking has always demanded music: you can't cook if you can't dance. In the beginning, the rooms had been under a spell of silence I couldn't break, as if I were a monk, abiding by the rules of an unseen abbot, but now I listened to my old favorites, the Cowboy Junkies, Dylan, Coltrane, as the pot on the stove simmered and the smell of something good replaced the caustic odors of paint and sawdust and the fire glowed in the woodstove.

There were days of putting down small filaments of routine as well. The voracious porcupine that cut the cable line and destroyed the Internet led me to the village library, an impressively pillared pile built in 1883 and named after a veteran of the War of 1812. It was a place to do research and get e-mail and be in the company of others. When school let out and the library got crowded I walked across the village green to the Whippletree Yarn Shop, and sat at a table and knit socks, and chatted with the owners: Shelley, just recovering from the end of a thirty-year marriage, and Andrea, whose giant schnauzer sometimes slept in the corner. Andrea in turn led me to Runamuck—I was looking for a place for Henry to play when I worked in the library, so he could be around other dogs, and Andrea directed me to a recently opened animal boarding and day care business just north of the village on Route 12.

Runamuck consisted of a shambling old house on the side of a hill, the right side of which was occupied by Cathy and John Peters and their two children, three-year-old Ondine and infant Amelia, and the left side of which belonged to an ever-changing assortment of dogs (first floor) and cats (top floor).

Out back was an enormous yard separated by a chain-link fence from grazing sheep and their guardian donkey, while the front of the house was dominated by the big yellow school bus John drove to make ends meet.

Henry took to the place immediately, and for a few days a week I had a schedule. I would take Henry to Runamuck early in the morning, talk to Cathy, a young woman with an imperturbable calm in the maelstrom of small children, barking dogs, spilled Cheerios, and ringing phones, or sit on the floor with Ondine—she was teaching me how to draw with crayons, and I was proving to be a very backward student. After that the day was mine, to go to the library and the yarn store, or back home to work, or to run errands, stopping off at the post office in South Woodstock to gossip with Jena the postmistress or to simply sit at one of the wooden tables in the South Woodstock Country Store and pretend to read while the regulars, big weathered men in thick plaid flannel, would discuss the prospects for the hay or the price of gas. Then at the end of the afternoon, I would make the rounds of the butcher and grocery store and the town's recycling center before heading back to Runamuck and a bounding welcome from Henry. The drive home against the cold, against the dark, made the prospect of homecoming all the sweeter—Halloween was drawing near, and there is no place spookier than New England at Halloween, especially in a place where the houses along the way are mostly dark and shuttered. (On some of the back roads, the inhabited houses were scarier than the empty ones—for a while I avoided taking one of my usual routes home after the only house in view put up a single decoration: a witch hanging from a homemade gallows.)

After dinner, right before bed, Henry went outside for his last walk of the day, and I accompanied him to take a final look at the night sky, to catch a glimpse of the distant silver crescent high in the sky to the east, or, if the time was right, at the fat full moon suspended low, surely close enough to touch. Bathed in that moonlight, I was content.

Looking back, of course, I see that I was recasting this new life in terms of the old—with trips to Runamuck substituting for my old routine of Zoë's school pickup and drop-off. No matter—the library, the post office, the yarn store, the doggy day care center—these were my first waypoints, the landmarks by which I could steer a course. Here were moorings—wobbly perhaps, but anchored—on a new map where I could knit myself, however tangentially, into the lives of others: watching Jena bloom after meeting a man she liked, or Allison grieving for her old dog, or commiserating with Holly, co-owner, with her brother Dan, of the general store, over the emotional ups and downs of our daughters' freshman trials and tribulations.

One morning I woke to a white world. It had snowed during the night, just a frosting, but the world was transformed, a dazzle of light and silvery branches and bitter cold. That morning my fingers stiffened and the sheets turned to cardboard on the clothesline and when I brought them in again they smelled of snow and pine. On the eaves, thick icicles had formed; I broke one off and gave it to Henry. He nearly keeled over with delight.

The snow brought its own difficulties—the ice-glazed snow was slippery, and it took a long time to find kindling dry enough to get the fire going; by the time I got back inside I was wet and full of complaint. And yet the ice-clotted trees, the cracking sound of breaking branches, the roaring wind and scudding clouds, were beautiful. Wasn't it better to be walking through slippery snow after some sticks of wood than to be trudging in the rain at rush hour to the corner store for milk? I am liking it here, I thought, a little startled.

A Gibbous Moon

Say goodbye to her, the Alexandria that is leaving.
Above all, don't fool yourself, don't say
it was a dream, your ears deceived you:
don't degrade yourself with empty hopes like these.

—C. P. CAVAFY, "The God Abandons Antony"

In November I went to New York for a long weekend. I had looked forward to the trip, to seeing Zoë most of all, but also to being in New York again, surrounded by the familiar.

I drove down a few days before her fall break started. At first the frantic city provided no comfort: the brazen storefronts with their lascivious mannequins, the billboards with their insistence that happiness consisted of tight clothes, expensive drinks, and women who would do anything to get them, were now nearly as foreign as lineaments for the nether regions of pigs. But gradually the culture shock faded and the knowledge that I was happy to be back asserted itself in a worrying sort of way. I had wanted not to like New York; I didn't want to be tempted away from Vermont.

And then Zoë arrived, and the little hive that had been my life buzzed back to life. The apartment was full of her friends and music and news, the days filled with the city's kaleidoscopic images: an art exhibit set in a pet store and featuring animatronic food; playing cribbage in the park while a lonesome fat guy wandered about with five green parrots on his head and shoulders and a poet pecked at a typewriter producing verses on demand; a young father holding his infant high in the air, smiling; the well-dressed wedding guests spilling out of a church on Christopher Street.

The last night of her visit, I had Zoë to myself. I had seen her only a few times since school began. In the early weeks of the semester she had called often, nearly every day, and nearly always when she was in crisis. Her class schedule would never work! First-year Russian and English symbolism conflicted— her academic aspirations were in flames! The drawing class was oversubscribed! The boys were drunken, fatheaded jocks! The girls wore sweatpants to class! I would give her the best advice I could and try to soothe her and then spend the rest of the day worrying. When I didn't hear from her, I was frantic, imagining her weeping into the red throw I had knitted. If the silence went on too long, heart in throat, I would call, and she would answer, sounding a little offhand, a little distracted, nothing like the hysterical child I had envisioned. Tentatively I would ask how she was. She was fine, she answered, as if this was an odd question, which for the sake of our long association she was willing to entertain. I mentioned the despair, the friendlessness, the lack of meaning in a meaningless world. Oh that, she said. Everything's fine. Could she call me later? She and six dozen pals were on their way to a party.

"You see, the good news is that they still need you," said my friend Susan. "The bad news is that they only want to talk to you when they need you."

Those calls had some benefit—I began to appreciate the distance. And sometimes, when I stood behind some highly strung stylish mother grappling with her rude seven-year-old in the local bookstore, wavering between polite admonishment for the benefit of onlookers and the urge to throttle the kid right there in line, I had to smile, glad to be exempt from at least some aspects of parenthood, to acknowledge that it was time for my daughter and me to go our separate ways.

But I forgot all that, lying next to her on the bed watching some dumb TV show we both loved. She was happier at school now—she had a close-knit group of friends, she was working on the school newspaper, she loved her classes. She was changing, growing up, and for the first time I had not been a day-to-day witness to the anecdotes and insights that were shaping her. The next morning she would be gone. I braced for the jolt of misery that should accompany that thought, but it didn't come. Instead, I thought of how, when I bent down to kiss her good night on her forehead, she still carried a faint trace of the scent that had been hers since the day she was born. I had not lost her. I never could.

I drove back to Vermont the next day and reached Castle Dismal late in the evening. The house was cold and dark. I groped for the wall switch. The house remained cold and dark. Something was wrong with the generator again. By candlelight I lit a fire in the woodstove and unrolled the sleeping bag and settled down to watch the flames and think about New York.

Yes, it had been lovely to wear nice clothes again and to watch the pageant of the streets, lovelier still to be back in the middle of things, to be connected, to be vital to someone. But I wasn't unhappy where I was, lying on the floor of a freezing cold house in the fitful firelight. There was something equally necessary here: the winter was coming, in the woods, in my life, a time when much happens out of sight. It would take some getting used to, but wasn't it possible that once I had, the perspective here at the margins would be as interesting as it was from the center, if not more so?

Growing old meant inevitable loss, yes, but that wasn't all it meant. Perhaps there wasn't anything to be done but to live through this time, to take possession of my grief, to claim sovereignty over my own sadness. One thing the country reminds you of forcefully is this: the darkness is as necessary as the light and must be met on its own terms.

Outside the window I caught the blurred outlines of a gibbous moon, but whether it was waxing or waning—and that night I very much wanted it to be waxing—I couldn't tell.

By the end of December, loneliness had begun to give way to a more comfortable solitude, the kind I had been hoping for, in which I was, for better or worse, most truly myself. In the fairy tales, in the old stories of wanderers and hermits, the forest was a secret, mysterious place of testing and transformation, where castoffs found their birthrights and wronged innocents their defenders. But there have always been those who repaired to the wilderness hoping for less miraculous results, who looked to Lenten seasons, untouched by daily distractions, as a way of coming to terms with themselves and their choices. Admiral Richard Byrd, the polar explorer, spent the winter of 1934 alone in an advanced weather base in the Antarctic. Apart from the scientific research he was doing, he wrote later, "I had no important purposes. There was nothing of that sort. Nothing whatever, except one man's desire to know that kind of experience to the full, to be by himself for a while and to taste peace and quiet and solitude long enough to know how good they really are."

In the oceanic expanse of the silence, the darkness and the harrowing beauty, Byrd found himself changing: "Yes, solitude is greater than I anticipated," he wrote in his journal, sixty-four days into the experience, "and many things which before were in solution in my mind now seem to be crystallizing. I am better able to tell what in the world is wheat to me and what is chaff."

I was beginning to have a small sense of what he meant. For me, solitude was roomy—it provided a space in which my half-formed assumptions about myself, the world, other people, unpacked themselves, stretched out and assumed their full shapes and walked about, giving me a chance to see them as they really were, and to assess them accordingly. So, too, with the cramped fears, regrets, and anxieties I was always too afraid to look at. Unpacked, no longer twisted in their ugly pretzeled shapes, they weren't so scary. In the light, in the quiet, we didn't get so much in each other's way.

Not always, of course: "There are days when solitude is a heady wine that intoxicates you with freedom," Colette observed, "others when it is a bitter tonic, and still others when it is a poison that makes you beat your head against the wall." There were still days when every love song on the supermarket Muzak lineup would make me cry, or I would find myself fantasizing about adopting the young schoolteacher sitting at the next table at the manicure salon because she hadn't had a date in seven years and her parents wouldn't let her live with them. But for the most part, I came to see loneliness as less of a fatal poison and more like a bad case of the flu—a temporary misery. I had found a tentative equanimity.

Or at least I thought I had.

Around that time, I had a visitor. Not a friend exactly, not an enemy certainly, but the holder of a title of dubious distinction—my last seducer.

Not my last lover (although they had been rather scarce on the ground for some time), but the last man who had held me in erotic thrall, the kind of scorched earth sexual obsession that leaves you burned and blistered and picking thorns out of your paw for years afterward without any regret whatsoever.

I had met him many years ago—one summer, when I was renting a place in Vermont, we had gone out for what I thought was a hike but turned into a masterful seduction, the like of which I had never encountered, involving a lost trail, an unexpected mountain lake, a stolen canoe, a vintage red wine, and a loon summoned out of the sky. If it had ended there, as I intended, it would have been a work of art, more beautiful than the *Winged Victory of Samothrace* or a roaring automobile, to borrow from the old Futurist manifesto, but, of course, it didn't end there and a few months later it shuddered to a standstill in pain, rage, and humiliation. I turned him into a monster, the predator, the vampire, the incubus, against whom I swam endless laps at the YMCA pool, trying to exorcise the hurt he had caused. I had not seen him for years, though he kept in touch in a desultory way, politely ignoring the cross and garlic I had laid in his path.

He was staying with friends nearby in New Hampshire, he said on the phone. Could he come for lunch? I was tempted to say no, as I did most times he called, but this time I hesitated. The day before I had been cleaning up the living room, listening to a favorite Lucinda Williams song in which the singer demands to know why she couldn't have both the pleasures of solitude and passionate kisses. And out of nowhere I was missing them, those passionate kisses and the mischief in a man's eyes, and so this time I said yes.

He drove up in a black Mercedes. I greeted him at the doorway. He looked old, my old seducer. He had been ill, and the illness showed in the deliberation of his walk and the translucent pallor of his skin. How hard the decline must have been for him, this man who reveled in his physical strength. His left eye wept, the side effect of recent surgery. He bore it all without comment, and his gallantry was touching. I had once seen a photograph from his youth, and in it he was beautiful, long-haired, bare-chested, muscles taut as he prepared to release the string on a hunting bow. I was overwhelmed now by how much of himself he had lost.

But it wasn't the physical changes that made him look strange to me. No, it was the fact that he looked out of place, the way a seal does when it leaves the water, the way any obsession does when it leaves the murk of your desire and your need and stands awkwardly in the light of day. *What was I thinking?* you ask yourself, so many years later. But of course thinking had had nothing to do with it.

He arrived carrying a big canvas bag. Inside was a bottle of wine, the same wine, he pointed out, we drank that summer afternoon so many years ago; I was surprised he remembered. I wondered what it would taste like untouched by the extravagant magic that had blessed everything that day. But he said he no longer drank wine, so I left it unopened on the mantelpiece.

I made him a good lunch. A roast chicken, a carrot ginger soup, a salad with blue cheese and pear and walnuts. We sat at the table and talked, a little awkwardly, about the economy, about the situation in the Middle East. And then the lunch was over.

He said he was cold. I went to the woodstove and picked up a log to throw on the fire, but when I straightened up he was right behind me. I turned, and he tried to kiss me and bumped instead into the rough bark of the wood between us. Wait, I said, surprised, because even though I had thought about it, must have expected it on some level, I had grown so used to the platonic nature of our infrequent dealings that I was taken by surprise. Or maybe what surprised me was how unmoved I was, how little I wanted what had once been oxygen to me.

Instead I made him a cup of tea, and he told me about the operations and the pain afterward and how bad the nightmares still were. He finished the tea quickly, anxious to get on the road before it got dark.

I watched him go.

That evening, I picked up a book, but the questions crowded in, claiming my attention. What story did I tell myself now about this thing that had brought so much pleasure and so much pain over the years? Were dusty memories enough to justify all the real estate this man and others had taken up inside my head? Perhaps it didn't matter now. Maybe it never had. And if so, what now?

Just this:

The night was cold, the fire warm, the dog slept while Chopin played. A glass of sherry glowed amber in the lamplight. I turned back to the page of the book I was reading. I told myself it was enough, but I was pretty sure I was lying.

Deal with Sex, I had written on the list I made when I first moved in, just ahead of Learn Latin.

The Latin part was easy enough to put into motion: I just needed to find a new copy of *Wheelock's Latin*, my favorite textbook on the subject. The other item wasn't so simple.

A woman has two jobs when she is young, said Simone de Beauvoir. One is to be a human being. The other is to be female. A balancing act, at best. One can drive you crazy. The other can save your life. Sometimes it's hard to know which is which.

You are young, and a light blinks on. A light that blinds you and dazzles you and makes you suddenly visible to yourself and to others. To men. You become something different in the light but you get used to it. And then, just as suddenly, the light goes out. And though you hated the glare, you grope for the switch.

In the years after Lee's death I realized I was becoming invisible in the world of men. I liked it sometimes. I hated it sometimes. This was curious country, an awkward place full of awkward questions. The ground beneath my feet shifted, and the world of things I thought I knew had become like a game of fifty-two pickup, with the exhilaration of flinging the cards— all of them!—into the air, the chastening realization that you are the one who has to pick them up again.

I would watch men and women together, sometimes with envy and sometimes with smug relief as they threaded their way through the quotidian clash of love and anger and the crowd that two can make. I am done with that, I told myself. But that wasn't quite true. I *wanted* to be done with that. Didn't I?

I found myself in a volatile state of change. My place in the world was up for grabs—freed, if that was the word, from the demands of fertility and confronted by the potential largesse of a life where sexuality played a very different role. I had always loved the power of desire, the way it affirmed my presence on the earth, the joyful thing it had been when I was young. But what could it be now that I was young no longer? At first I was frightened, sad, and very angry.

I remembered something Isak Dinesen once said: "In Africa, all old women had the consolation of witchcraft; their relations with witchcraft were comparable to their relations to the art of seduction. One cannot understand how we, who will have nothing to do with witchcraft, can bear to grow old."

And yet sometimes, walking home through Washington Square Park in the afternoon in those first years after Lee's death, I felt a buoyancy, a lightness I had not known in a very long time. I looked at the young women in their twenties sauntering down the street, their carefully made-up faces composed into masks of indifference to the attention they courted and disdained; I looked at the couples entwined on the benches, rapt with the drama only they could see, at the mothers in the park hoisting children onto swings, their eyes anxious, bored, amused, weary. I've been to each of those places, I thought, and now, for the first time in a long time, I was in a place I'd never been before. Who was this woman who was no longer lashed so tightly to the world of men; what did it mean to be finally getting old, to live alone, to be invisible in a way that I had not been since I was a teenager? I was nervous, but I was excited as well.

"You only begin to discover the difference between what you really are, your real self and your appearance when you get a bit older," Doris Lessing said in an interview in *Harper's* in 1973. "A whole dimension of life suddenly slides away and you realize that what in fact you've been using to get attention has been what you look like. . . . It's a biological thing. It's totally and absolutely impersonal. It really is a most salutary and fascinating thing to go through, shedding it all. Growing old is really extraordinarily interesting."

Lessing was British by way of Zimbabwe: perhaps that accounts for the crisp understatement. Salutary? Interesting? Yes, the way a dive into ice-cold water is interesting. Leaving that kind of desirability behind is also scary, bewildering, and disorienting. Still, Lessing's approach was much more comforting than contemporary commentary on the subject, books that talked about "juicy crones" and "seasoned women" (culinary metaphors were apparently de rigueur where older women were concerned) and seemed to mandate that to be really happy, women must remain as randy as rabbits as they tottered toward the grave.

Which was unfortunate, because the choice was not entirely mine to make. What lay ahead was an age, in Judith Thurman's elegant phrase, "of increased authority, erotic exigency and forced retirement."

Besides, I liked the idea of retirement, of thinking of sex as just a handful of seasons in one's life—the sense that redemption lies in knowing when to leave it all behind.

Desire, however, doesn't have the good grace to die, simply

because you ask it politely to do so. I was angry with myself for still wanting love, for still craving all that went with it. It was an embarrassment, a barnacle, something I should have outgrown. Besides, I didn't know what I wanted from sex anymore.

When I was young, desire had been a drug, one that I wanted for much the same reasons I would court getting lost—because it let you off the hook from your ordinary life, the one where I was always late, always lacking. I loved the fever of it, the color in which it drenched everything, the excitement with which it imbued the act of turning a corner, answering the phone, checking e-mail. Desire had been my theater of war, my coming-of-age, the weapon of choice in my own rebellion. And now?

I wasn't sure. I was no longer at an age where I simply saw what I wanted to see in the eyes of the other; I couldn't pretzel myself into the person I thought he wanted. Besides, I knew what love was now; it was hard to settle for less.

Still, I couldn't accept that love was over. I started to make bargains with myself, the way you do when you're quitting cigarettes or anything you truly love. One more affair, I'd think to myself. After that, the veil.

Finally it got to the point where I could not for one minute pretend I didn't miss the touch of flesh against flesh. I did something I had sworn I would never do and created a profile on a popular Internet dating site. Before long I was driving over the George Washington Bridge to have dinner with a good-looking stranger who advertised himself as an entrepreneur, whatever that was, and who had been attracted to a line in my ad that said I was looking for a pirate who had learned a thing or two.

The stranger was handsome, affable, and not very bright. The first night we had dinner and a long lingering kiss that told me what his conversation had not, that I wanted to see him again. A few dates later, he took me to his apartment and I had to laugh. It was decorated in a style I knew well: mattress on the bare floor; flamboyant, expensive quasi-erotic paintings and sculpture collected from souks in Marrakcsh and Istanbul; a set of scales tucked back neatly on the kitchen shelf; a couple of burn marks on the battered coffee table; and an unmistakable aroma embedded deep into the upholstery. Ah, so you're a drug dealer, I said.

Yup, he said. How did you know?

Just a guess.

The next morning I rose early from a rumpled bed where a man lay sleeping, my eyes scratchy from lack of sleep, remembering how much I loved this leaving of a strange bed, of being the one that gets to go.

Colette wrote often about the degrees of satisfaction a man can afford, "from a good meal to a solid mystical engagement." The former, she thought, was not to be disdained: "a nice little nothing well-presented is already something."

I thought about the man whose bed I left that morning. He was none of these things a lover had once been to me, not God, not the devil, not friend or enemy, not poetry or prose. Making love to him had been the nice little nothing Colette had described it to be. I played with the notion that I would see him again, wondering if I was now at a stage where sex could simply be great good fun, like swimming, like getting stoned, something as ephemeral as blowing bubbles on a windy day.

But I knew already that it could never be like that, not for me. The evening had proved what I already knew, that desire detonated huge holes in contentment, and if it didn't, well, then, I wasn't much interested.

I had read somewhere that the ancient Greeks treated erotic desire like a flu, and sent condolence cards to any friend who had the misfortune of contracting a bad case of it. That seemed about right. Desire was not a thing to be encouraged. In Vermont, I had decided, I would give up sex altogether—I was afraid of losing any more of myself than I had already, and I had never solved the question of how to be myself and be in love with anyone but Lee.

And it had worked, for a while. But the meeting with my old friend had shaken me, rattled the bones of a question I had come to Vermont to put to rest, or, perhaps, through which to drive a stake. He left behind a longing I couldn't dismiss.

Wheelock was still in print, I was glad to discover, going strong in its seventh edition. I ordered a copy online, and while I was at it, posted a profile on Match.com.

What's This?

NEW IN TOWN" read the headline on my profile. What followed was some mortifying nonsense that tried to sound confident but not intimidating, vulnerable but not needy, witty but not brittle. I probably shouldn't have worried so much—according to some research, men looking for women on Internet dating sites don't usually read beyond the headline, once the picture passes muster. My approach was essentially passive: I didn't bother looking at the prospective candidates in any detail. Former experience had proved that men in my age group tended to sound and look a lot like puppies left out in the rain. So I lowered my traps in the water and waited.

One of the advantages of living in a state with a very small population is that you don't encounter the same competition you do in a place like New York. Before long a message came sailing in from an age-appropriate candidate who wrote well, seemed to be more or less sane, and possessed of a sense of humor. Let's call him Mitch. Sometime before Christmas, Mitch and I started up a fairly steady e-mail correspondence—he was funny, personable, self-deprecating, and showed no inclination to wax rhapsodic about holding hands on the beach or similar Internet clichés.

We met the first time at my house. He was good-looking enough, tall, long-legged, and reasonably fit, with gray-blue eyes set in a kindly nest of lines. He was mostly bald, it turned out, once he mustered up the courage to take off his red and blue ski cap. (Hand knit, I noticed. Ex-wife? Girlfriend?)

All in all a perfectly presentable package, but one that quickened nothing in body or soul. Maybe it was the eyebrows, on which for some reason I fixated instantly. They were like furry little hyphens, very short and straight across, that bobbed up and down on the broad expanse of his forehead as he spoke, lending him a slightly sheepish expression. It was a bland face that looked like one of those fonts used in a child's first chapter book.

We took a long walk up Wild Apple Road, one of my favorites, and it was beautiful, flanked by rolling meadows of snow stretching out to the hills that smudged the brilliant blue horizon.

Afterward, he opened up a bottle of cheap red wine and I warmed up a stew, and we talked and talked, or rather he did, nattering on about his construction business, about how he had been a disappointment to his father (I am surprised, and moved, in my intermittent forays into middle-aged Internet dating, how many grown men not only bear the scars of their upbringing but also want and need to talk about them), and more tellingly, about his ex-girlfriend. She was overweight and unattractive, he said, and she had three small children, but she was his best friend and . . .

He went on and on, and I began to want him gone, tired of party manners, bright and sprightly and artificial—"Are you drunk?" Zoë asked when I answered the phone, acutely aware of the man in the room. But sitting there on the sofa next to him, another conversation had begun. His physical presence asserted itself, and perhaps it was the novelty of being alone in a room with a man for the first time in a long while, but I felt an excitement stirring. Perhaps we'd make out a little and something would kindle. But no—at nine-thirty he left. I could say something that would embarrass us both, he said, before kissing me very lightly on the lips.

The next afternoon a polite bread-and-butter e-mail arrived, and then nothing more for several days. I felt a rejection out of all proportion to the tepid interest Mitch had aroused. The weather was bleak, cheerless, and I went out for long walks in heavy wet snow with Henry, returning to the house chilled, unwashed, and grubby in my thick layers of snow clothes but lacking the initiative even to take a warm bath, as if the leaden clouds pressing on the house had settled in my soul. I tried to work, but I was foundering, slow and uninspired. I had to find a way to feel some urgency, to shake off the torpor. The small gains I thought I had made disappeared so easily; I couldn't remember what they were. At least, I thought, I'm not in New York.

Mitch and I made a plan to meet again, this time at his place. A week later, on a Saturday afternoon in February, I found myself driving up an icy mountain road while Henry sat at red alert in backseat-driver mode, his hackles raised, watching the road and, I was convinced, muttering to himself, "Slow down! Go faster! Pass that guy! Downshift, you moron"— through two hours of hairpin turns and iced-over bridges.

We arrived finally at a sleek modern house perched at the edge of a bright field of snow under a bright blue bowl of a sky and decanted ourselves into the driveway. Mitch emerged with a young springer spaniel at his heels, and the two dogs bounded off across the snowy fields. There was the obligatory tour of the house, which, like the spaniel, turned out to belong to his ex-wife, for whom the object of my—well, I wasn't sure what he was the object of—was house-sitting. Then a long walk during which he told me about his reckless youth, skating for hours on acid, the years bumming around Morocco, the 110 mph drives to Boston on one harebrained scheme after another—and dear God, he began to sound more interesting. Even the unfortunate eyebrows weren't so bothersome. Was I just molding him, pretzel-like, into an archetype that had always attracted me? Was I talking myself into this because I was afraid I'd lost the capacity to be attracted (or attractive) to anyone?

The answers were yes and yes. In search-and-rescue circles, this behavior is known as route delusion. It's what lost persons do when they've been walking in circles for hours, ending up at the very place they were trying to leave.

Back at the house we ate a perfectly decent lentil stew— I'm not sure what it says about my dating choices, but I've yet to go out with a man who doesn't have a lentil stew in his repertoire—followed by thin slices of Asiago cheese and smoked Gouda. When it was time to go, Mitch gave me a box of kindling as a parting gift and walked me to the car, where he kissed me. A less-than-full-throttle kiss, but more than a peck this time, and then another one, and I began to feel something like the old champagne bubbling up.

I kissed him one last time and then drove down the road, very pleased with him and with myself, wondering how soon I would see him again and how soon we would make love. I began to settle myself into a happy fantasy of the things we would do and the places we might go. Yes, I admitted to myself, I had missed a man in my life and he was just what I needed: nothing like Lee, no one I could fall in love with, and yet a possible companion. All of this, despite the less-than-captivating conversations, the not-quite kisses, the lurking ex-girlfriend.

The champagne lasted for a day or two, until an e-mail arrived. "About those kisses. A bit confused I am by those three tantalizing kisses . . . Actually I'm confused not a bit about the kisses (they were just fine) but more about my head and where it is at presently. I seem to have trouble with being in the present. Will cogitate and sort and like the machine that filters and wraps spare change so tidily, I'll try and make sense out of what's rumbling around in my spare pockets of the cerebellum. I had a wonderful skate during lunch yesterday and intend to do the same today. . . ."

I read the e-mail and wrote back, thanking him for his honesty and assuring him that I understood the sensitive nature of his feelings and would be glad, whatever the outcome of his hesitation, that we had met. Didn't mean a word of it. There followed in his next e-mail a dithery string of ambivalence—yes, the old girlfriend was still in his heart, but gee, he wasn't really attracted to her, but yes, he felt an obligation, but never mind, what the hell, let's do this thing. He would come to dinner that weekend, after I got back from New York, where I was headed for a few days for a round of long-overdue medical and dental checkups.

What's this?"

It was a beautiful day. The city blazed with a hard metallic sheen of bright sunshine and bitter cold. The shrink and I had had a lively conversation on one of the less tedious tropes in my psychological canon and the dental hygienist had praised my excellent flossing. What else was there? Who needed love? I was fine, just the way I was.

Not so fast, said the Kindly Ones.

Ron Ruden had been my doctor since I first moved to New York. We had daughters exactly the same age, and it was he, more than the Olympian and unreachable doctors at Sloan-Kettering, who had guided me through my husband's illness. My annual visits to his office were mainly spent catching up on family news, with a mere ten minutes devoted to the requisite poking and prodding and drawing of blood.

So my mind was far away, planning that night's menu, when his question brought me back to the reality of the exam room. He was palpating my left breast.

What's what? I asked.

This, he said. He guided my hand to something hard near the breastbone, which in fact was what I had thought it was when I had done my own manual examinations.

It was a mass. A large one. Dr. Ruden was talking but I couldn't seem to hear him. I got dressed while he called the hospital and scheduled an emergency mammogram for the next day.

Back in his office, he said, It's probably nothing.

I looked straight into his eyes and read what was written there.

Dr. Ruden? You don't play poker, do you?

No.

Good.

It was cancer, but it could have been anything, really. The predictable life markers, the milestones on the well-trodden path —the graduations, the promotions, the anniversaries— are more than enough to cope with, but then there are the unholy catastrophes, the curveballs you don't see coming— the layoffs, the illnesses, the parent who can no longer find her way home. What the experts call survival situations. "It's easy to imagine," wrote Laurence Gonzales in *Deep Survival: Who Lives, Who Dies, and Why,* that such situations "would involve equipment, training, and experience. It turns out that, at the moment of truth, those might be good things to have but they aren't decisive. . . . In fact that experience, training, and modern equipment can betray you. The maddening thing for someone with a Western scientific turn of mind is that it's not what's in your pack that separates the quick from the dead. It's not even what's in your mind. Corny as it sounds, it's what's in your heart."

That night, I called Zoë. She burst into tears but quickly

recovered. I told her this was nothing, that comparing my situation to her father's was like comparing the sniffles to tuberculosis, that I needed her strong, stoic, and black-humored self by my side. Then I told her the joke a character in a short story by Lorrie Moore tells, one that had always amused Lee. A man is told by his doctor that he has six weeks to live. "I want a second opinion," he says. "OK," the doctor says. "You're ugly, too."

Alexander Swistel, the surgeon Dr. Ruden had recommended, couldn't see me for a week or two, so I went back to Vermont and into a limbo of shock, anxiety, nausea, and a whirlwind of emotions so strong, so operatic in their intensity, that they rendered the mundane concerns of everyday life alluring but unreachable, like an amusement park on the opposite side of an unfordable stream.

Which is why, I suppose, I decided not to cancel the dinner with Mitch. You would think that a potentially fatal illness is just the thing to put an abortive romance with a middle-aged slacker into perspective. But if romance has always been your drug of choice, the thing you turned to in good times and bad, then it will get you to oblivion as well as any other.

Besides, he had been wonderful when I told him the news over the phone. I returned to Castle Dismal to find him waiting, a warm fire blazing, the refrigerator stocked. We took a long walk, and he was kind and undaunted about the future. Are you sure you want to go out with a bald-headed woman? I asked, not wanting to mention the more permanent physical changes in store. I love bald-headed women! he said, and I felt a little better. Perhaps I wasn't going off a cliff after all. He left after that, promising to come back for dinner over the weekend.

I worked hard in the days before the dinner, cleaning up the house, making a complicated lamb tagine, driving miles to find fresh flowers for the table. Mitch was going to let me know the night before what time to expect him. But the evening came, and the evening went, and the e-mail I sent inquiring after his whereabouts went unanswered.

The next day he wrote back: I'm really really sorry, it began. The world is moving very fast and I'm having a little difficulty keeping up so no, I can't come down today but will call this evening. Talk to you later and have a good day . . . got to run out the door!

What had happened, he told me in the phone call that eventually followed, was simple enough. The ex-girlfriend wanted him back. They had talked, and there was still something there.

I took it hard. I would like to think that it was the cancer talking, but I probably would have felt the same way without the diagnosis, given how hard a time I had been having with the prospect of being on my own forever. Either way, I was chagrined at how much I had wanted this to happen, how easily all my foolhardy, noble notions of forswearing men for all time had sailed out the window, how differently I had felt driving back home from Mitch's place after one afternoon and three anemic kisses, as if I'd been somehow validated. And when it was stripped away, I felt less whole.

I threw out the lamb tagine and indulged in a mawkish, angry evening. First I reread all of Mitch's e-mails, and reveled in their banality. Then I listened to Leonard Cohen in the dark with a glass of sherry. Early Cohen is perfect when you want to remember just what arrogant twits men can be. (Late Cohen, after his voice changed and he had learned a thing or two, is better with a dry martini and a cup of rueful reminiscence about what twits we all can be.)

The next morning, I felt much better—it is relatively easy to get over someone you had talked yourself into liking in the first place—but nevertheless I went through the usual cathartic ritual: the purging of the phone numbers and e-mail address from the contacts list, the deletion of the correspondence from the computer. I canceled the Match.com account. Then I wrote two e-mails to Mitch. One was noble, calm, and generous, if just a wee bit guilt trippy; the other was a wicked, mean, vindictive screed that made me laugh when I read it over. I sent the first and kept the second to cheer me up whenever the cancer threatened to overwhelm me.

By then it had been a week since the diagnosis and I had become three people. One was a disinterested observer, watching a colossal train wreck and thinking to herself, Oh dear, those poor people: that was the writer, taking notes. The second was a kind of protean emotional swamp creature who morphed from a frightened field mouse into a gallant (and I wished this wouldn't soon become an all-too-appropriate description) Amazonian warrior, ready to stare down any enemy, back into a trembling bit of fluff in less than a minute.

She in turn yielded to my own personal manifestation of Kali, the Hindu goddess of death and destruction, black with rage, bearing a skull-topped staff and wearing a garland of heads, Kali of the gaping mouth and lolling tongue and fiery red eyes, filling the sky with her roars, devouring her victims, dancing on the corpses of her enemies. I would drive down the highway, the snow swirling white along the blacktop, possessed by a kind of exhilaration—I can't figure out another word for it—at the idea of leaving my old world in ashes. Yes, cut off the breast, renounce sex and love and womanhood and contentment, revel in chaos, rejoice in a wholly impersonal universe, where the whirling snow and the injured body and the thin air of catastrophe were one and the same and equally meaningless. Off with their heads! shrieked the Red Queen. Off with their heads! And then I understood the dark power of the old crones in the legends and the fairy tales, the women who had lost everything but the rage that kept them alive.

I began to tell people my news, each announcement sounding like some ridiculous lie—why on earth would I say such a thing? People reacted as people do, according to their nature. What both touched me and nettled me was the nearly universal need to make it better, to fix it, if only for the moment, to soften the blow. You'll be fine! they all said. You have to be brave! My cousin had it and now she's running marathons! It could be worse!

The reactions brought my relationship with Lee to mind, how at first, when I came to him with bruised feelings, or disappointed hopes or fears for the future, he would set about fixing me, telling me what I should do and how I could overcome whatever obstacle had flummoxed me, with the enthusiasm of an energetic border collie determined to chivvy me back to the herd of the contented and confident. Sometimes, though, all I wanted were four short words and an accompanying tender glance: *you poor sweet baby.* Eventually, we got it down to a formula, as couples do after a while. I would cry or complain, he would start in on the border collie, and I would hold up four fingers.

Oh right! he would say. You poor sweet baby! in so triumphant a tone that I always had to laugh.

My favorite response to my situation, however, came from my friend Lynne, whose blunt Yankee pragmatism had always served to protect the easily aroused sensitivity that lay beneath it. I told her that I would probably need a mastectomy. Well, she said, casting an appraising glance at my seated form, the good news is, your breasts are small. You'll hardly be able to tell the difference.

Others were more overtly comforting. I blurted out the news to Harriet in the parking lot at Runamuck. She thought for a second. My best friend died of breast cancer this year, she said.

I raised an eyebrow. And this is supposed to make me feel better how? I asked.

It's not, Harriet said. But my friend was eighty years old when the breast cancer came back. The first time she had cancer she was sixty. That meant she had twenty good years in between. I think that's pretty good.

I would have hugged her then. Luckily it was out of the question—Harriet was a New Englander born and bred. But I was grateful and, for the first time, optimistic. Harriet had provided a way marker, pointing in the direction I needed to go. It wasn't sympathy I needed, or encouragement. It was a clear eye and a steady hand on the tiller, a realistic sense of what I was up against. A map.

One afternoon there was a knock at the door. I looked out the window—no car in the driveway. That meant only one

person, and yes, it was Tom, my next-door neighbor, already
in the mudroom, calling my name. I braced for battle. I didn't
care what harebrained flatlander thing I'd done this time, I
didn't want to hear about it and was prepared to tell him that
I had just about enough on my plate, thank you very much,
without another ear beating, as my grandmother would have
called it, from him. I was channeling that formidable woman
when Tom appeared in the kitchen, holding a delicate blue
bowl, a snow-white cloth napkin, and a mason jar full of something that smelled delicious.

My wife made it for you, he said. I had met Catharine, or
Cat as she was called, on one of her visits to Woodstock. She
didn't come often: the Vermont house was Tom's retreat, where
he could play his music and have time to himself. She was a
warm woman with an incandescent smile and a down-to-earth
wit. It's her Portuguese kale soup, Tom explained. It's really
good for you, and you need to eat healthy. I'm sorry about the,
about the—he could barely bring himself to say the word.

Cancer? I said. I thanked him for the soup.

If there was anything he could do, he said, anything at all,
just call—he would walk the dog or bring in the firewood, anything.

That was when I knew things were really, really bad. I was
so touched by his kindness that my eyes began to fill, but Tom
knew how to avoid a scene. "Look at it this way," he said. "At
least you won't have to worry about dating anymore."

Alexander Swistel was a dapper man, inclined toward bow
ties and a suavity that he wore with proper Ivy League insouciance. Well, he said after glancing at my left breast, which the

biopsy had left looking like it had gone five rounds with a washing machine. You certainly didn't give me much to work with.

You're the first man who ever complained, I said.

Given the size of the tumor, Swistel said, he would not operate right away. He suggested I have chemotherapy first in the hopes of shrinking the thing and thereby salvaging the chance of a lumpectomy. I left his office mildly encouraged.

Bonnie Reichman, the oncologist Swistel recommended, took a different tack. Part of it was a difference in personalities: Dr. Swistel discussed treatment options with a reassuring offhandedness, as if we were discussing the game plan for next week's office party, but then he could afford to: he did most of his work while his patient was unconscious. Which is not to belittle his reputation—he has been a pioneer in painstaking surgical methods that preserve as much of a patient's normal appearance as possible; others in the field describe him as an artist. But Dr. Reichman shepherded her patients through the nerve-racking, debilitating hairpin turns of a treatment that can last every day for years. She was the one who had to tell a woman her future, when there was no future left.

Yes, she said, chemo might shrink the tumor. It might not. More likely I would have six rounds of chemo, surgery, probably mastectomy, radiation, and possibly more chemo. In addition there would be a year of targeted therapy delivered intravenously. And after that, five years of hormone therapy, which could turn bones to glass, increase weight and weaken the heart, and set the stage for other cancers. Or not. I was one of the lucky ones, she said without any irony: the cancer had been caught at a relatively early stage, I was otherwise healthy, and there was no sign of metastasis.

I knew she was right, but I walked out stunned and, I'm afraid, in tears. The next few days were taken up with the usual cancer treadmill—a raft of tests to make sure my physical self could survive the treatment, and then a round of shopping for hats and a wig, to make sure my ego would. The wigs all looked like roadkill; it was impossible to imagine wearing any of them. Then I drove back to Vermont, with what looked like a dead muskrat in my suitcase, and a small mountain of pamphlets about everything from what to eat during chemotherapy to how to pencil in eyebrows when I had none. Coming down the driveway I stopped and sat in the car and listened to the wind in the trees for a long, long time. Everything looked the same. And nothing was.

In 1971, a lightning-struck airplane fell into the Peruvian jungle. One of the passengers was a seventeen-year-old girl. Disobeying all of the standard advice for being stranded in such a place, she did not stay put and wait for rescue, as the other surviving passengers did. Instead, she started walking, still wearing her white confirmation dress and a pair of high heels.

Eleven days later, having made her way through some of the densest jungle on earth, she arrived at an empty hut on the banks of a river, and collapsed. She was starving, dehydrated, and covered in leeches and worms erupting from eggs laid under her skin. She was found by three hunters who had happened by and took her back to civilization.

The other surviving passengers had stayed where they were, awaiting rescue, which is exactly what any survival guide would have advised. Besides, they were convinced they would never make it out of the jungle alive. And none of them did. The girl had taken one look at the jungle canopy and decided no plane would ever see them, which in fact was the case.

The difference came down to personality in the end. The teenager, like most of her adolescent tribe, wasn't big on patience and probably shared their inflated notions of her own immortality. The others followed the rules that maturity had taught them. There was no wrong choice. It was a crapshoot—the outcome of any disaster is a crapshoot—all you can control are the decisions you make, what eccentric combination of logic, intuition, personality, and pragmatism leads you to a kind of hope.

In the pink-ribbony dream world of breast cancer, the jungle canopy is composed of information. The Internet is thick with it, the reliable and the utterly specious, the frightening and the inspiring. The books are plentiful as well, but it's the words *I looked up breast cancer on the Web and . . .* that drive the cancer specialists crazy. Dr. Reichman gave me the name of one site she considered reliable, and urged me to stick to that. Of course I ignored her. I bought nearly every book I could find, from the encyclopedias of doom to the unctuous purveyors of treacly optimism. I spent hundreds of hours on the net, unable to stop myself, reading everything I could find.

The obvious disconnect between the grim statistics, the unequivocal fact that there was no cure for this disease, and the pie-eyed optimism of the Web sites drove me nuts. The pink ribbon brigades never met a side effect they couldn't minimize and yet insisted on reminding you that your life would never, ever be the same. I felt condescended to and intensely frustrated, and all the encouraging words made me more afraid—was the unadorned reality of breast cancer really that scary?

I tried to prepare myself for whatever it was the cheerleaders weren't telling me. I googled images of bald women and graphic pictures of mastectomies and breast reconstructions—the good, the bad, and the grotesque. I read blogs by the brave and the terrified, discussion boards on topics ranging from hair loss to horror stories of insensitive husbands, evil insurance companies, and side effects of Gothic dimensions. I read arcane scientific journals that I tried to translate from languages I didn't even speak.

Cancer porn is what Catherine Lord called this sick fascination with every detail and every disaster story in *The Summer of Her Baldness*, my favorite breast cancer memoir. It's a perfect description; the stuff is lurid, often amateurish, and utterly hypnotic.

There are as many ways of coping with breast cancer as there are patients, and there is no preferred method as far as survival goes. Total denial seems to work as well as the diligent industry of those who spend hours imagining golden armies of healthy cells battling evil malignant ones. Some women watch their diets. Others live on chocolate milk shakes.

But I was convinced that if I just kept reading I would eventually find the thing that was missing, the piece of information that would give me the traction I was lacking.

My need for information was the legacy no doubt of too many years as a journalist, the sense that there was never such a thing as too many facts. Somewhere out there was the one statistic or piece of research that would provide a landmark by which to steer a course. Without it, I cartwheeled between shock and denial and depression.

I found it finally on a Web site called, simply enough, Cancer Monthly. In the section devoted to breast cancer I read:

> Nearly half of all patients who are treated for apparently localized breast cancer develop metastatic disease. And half of all initial cancer recurrences occur more than five years after initial therapy. Although a very small number of these patients can enjoy long remissions when treated with combinations of systemic and local therapy, most eventually succumb to their cancer.

There it was: the elephant in the room, the thing everyone danced around and no one wanted to talk about. Is it true? I asked Dr. Reichman. She gave me the standard answer to questions concerning mortality rates. Would it make any difference if it was?

It's a good question, because the answer is nearly always no—the patient is going to do whatever she can to get better and live longer, and recognizing that fact can help quell the anxiety at least a little.

That was my answer as well. But I still needed to know.

Yes, Reichman said, it was true more or less, and after that I calmed down. I felt more in control, if only of the facts. For me, there was some peace in seeing spelled out in print what the word *incurable* makes fairly clear: most women with breast cancer will eventually die of it if something else doesn't get them first, whether that something is a car wreck or old age.

The experts will tell you that survival in the wilderness or even in the shopping mall, if that's where you happen to be lost, depends on optimism and pragmatism, on a healthy dose of hope and the ability to read the altered circumstances you confront with some degree of accuracy. But then there are the cases, like the Peruvian plane crash, where about all you have is an attitude, not the odds, but a way of coping with the odds: to face death walking out of the jungle or wait for it to come to you. I think that whatever understanding I gained about direction, about orienting myself in an unfamiliar world, began in earnest then, when I learned how to face the present without blinking and decide for myself how I would cope.

Chemo would not be so bad if it were not for the insects, such ugly things, with their sharp mandibles and terrible teeth and pincers that tunnel so deeply, relentlessly into the brain. I had not known that insects could be made of metal. The beetles are big and black and they dig their way into the nightmare place; I can't remember in which part of the brain that's located. Medulla oblongata? I should have studied more. If only there weren't so many of them, if only their bodies didn't rub against one another, making that terrible scritch, screech, scraping sound, like the lawn mower when I haven't oiled its blades. Why won't they stop? If they did, then I could sleep through these endless hours while delicate insect legs skitter across the soft tissues of my brain and insect talons work away. I have been on this sofa, unable to move, for hours. I look at the clock. How long has it been since last I looked? Five minutes.

The third day of chemo was the worst, a gray shadowy pit of time full of nightmares and fever dreams. Then the days gradually got better until the third week, when I would feel nearly normal, when it was time to do it again. I had the chemo in Dr. Reichman's office in Midtown Manhattan. Having seen other chemotherapy delivery rooms when Lee was ill, I knew how lucky I was. Dr. Reichman had set up hers more along the lines of a cozy small-town beauty parlor than a grim factory for pouring lethal chemicals into fragile human containers.

There were only three chairs in what I came to think of as the chemo salon, leather or at least leatherlike recliners of a size and magnitude that demanded either a huge helmeted hair dryer hovering above or a basin for pedicures placed at the foot. On one side of each chair stood a small table for water bottles and other necessities and on the other a wheeled metal dispenser from which the bags of chemicals could be suspended while they dripped through plastic tubing into a needle taped securely to a forearm.

A well-stocked folder of take-out menus was in reach next to a pile of magazines and extra blankets and pillows, as well as DVDs of movies that could be popped into laptop computers. Tucked into one corner of the room was a kitchenette with a small refrigerator filled with bottles of water, juice, and diet soft drinks, a sink, and, on the counter, a bowl of healthy low-calorie snacks—granola bars, hundred-calorie cookie bags, dried fruit.

There were a few chairs placed for friends and family as well, but most women came alone. The chemo took about four to six hours to deliver, depending on the particular cocktail prescribed, so there was time to talk, and over the months we got to know one another a little. The patients varied in age and income but also in how many times they had had cancer and how advanced it was. First-timers were probably the majority, and most of us had fairly good prognoses, but there were patients who were back with a recurrence, or a metastasis.

There were lawyers, and housewives, and grandmothers; there was a woman in her seventies who was a psychoanalyst by day and a cabaret singer by night, who had chemo four times a week and would do so for the rest of her life: the cancer had spread to nearly every organ in her body. She would spend the time going over musical scores with her husband/manager for that evening's performance. We were all in awe of her.

Those of us going through our first bout with cancer tended to be a chatty bunch, especially at the beginning of the session some of us arrived slightly hyper from the steroids taken the night before to keep the nausea at bay. We spent a lot of time discussing surgical options and side effects with the gusto of a convention of car mechanics or tax experts or any group possessed of arcane knowledge of passionate interest only to themselves. But after the intravenous Benadryl began to take effect, we were all pretty stoned, and the chemo salon took on the relaxed, slightly louche atmosphere of a high-class opium den. Sometimes, in the background, we could hear, as if in a dream, Dr. Reichman's low, patiently insistent voice trying to convince a frightened woman to continue treatment despite the side effects, or explaining the options left to someone for whom they had contracted to very few.

I grew to love these women, even the ones who couldn't seem to move a muscle to help themselves, especially the ones who couldn't, who needed more blankets and more water and more attention than anyone else; their anxiety spoke for all of us. They tended to be the older ones, the last remnant of the generation that had depended on husbands and fathers to make everything better; they had never had to cope by themselves. It was much harder to know what to say to the younger women, because of the monstrous unfairness of it, women who hadn't even had time to have children or had two small ones at home—and Jane, our oncological nurse, said the patients were getting younger all the time.

I would drive back to Castle Dismal a day or two after the infusion. On the bad days I would rest in bed and stare out the big picture window in the bedroom at the bare trees, fixing in particular on a big birch that stood out in its white bare perfection from the dun-colored branches of the others, staring at it as if the strong sturdy pillar of its beauty could somehow anchor me, if only I held on tight.

That spring was my first mud season in Vermont. Some mornings after it had rained all night, I woke up to murky waters rising, puddles pooling, torrents reaching up to Henry's flanks when he ran outside to see what was going on. The water rushed down the road from the woods, taking out most of the driveway. The floor and the furniture were patterned in muddy paw prints, and my clothes were caked with dirt. I was fascinated by this wet and rushing violent change, the muck and ooze, the motion of it, water and road and wind and rain, the loud smacking, streaming, sucking, crashing, moving, fall-

ing, melting mass of it. If you had to be in the middle of a catastrophe, then I was in the right place, because spring in New
England is a catastrophic season.

One night, a couple of weeks after the first infusion, I absently
ran a hand through my hair and a few strands came away in
my fingers.

Such an odd sensation, the slight tug as the scalp released
its hold on the roots, an instant of realization of what is happening and then the visual evidence, there in your hand. I had
tried to prepare for the hair loss—watching Internet videos on
creative head scarf tying, buying yards of fabrics in ridiculous
colors, practicing with the free makeup given out by the American Cancer Society at one of its "Look Good Feel Better" sessions, in which volunteers demonstrate the use of foundation
and blush to look less sick. But I don't think I ever believed it
would happen.

When it did, I headed to the yarn shop for a pattern and
some wool to make myself a hat. Something that wouldn't
make me look too cancery, I said to Karen, the young woman
working the desk that day. But the women of Whippletree were
way ahead of me: they had designed a hat—the Darling Beret,
they called it—and Karen had knitted it in a soft mauve cotton
with a couple of darker stripes around the brim. It was beautiful and perfect and I had no words for their kindness.

After the first round I had waited nervously for the first sign
of the coming change, tugging nervously at my head about
fifty times a day. I had read the discussion boards about how
traumatic it was to lose your hair, as well as the psychologi-

cal studies, in which many breast cancer patients claimed that losing their hair was much more upsetting than losing their breast.

But while it was strange when it actually happened, my re-action, past the first little rush of fear, was relief: Now it begins, I thought. Now there is something I can actually do, beyond waiting in dread.

It was late at night. I went upstairs to the bathroom, put a towel around my shoulders, and in the clinical glare of the flu-orescent light, I cut my hair as close to the scalp as I could with a pair of scissors. I had read somewhere that it was better to get rid of your hair once it began to fall out, much better than waking up every morning with the evidence all over the pillow.

So I cut it off in hanks, laying the fistfuls of brown hair on the linoleum counter. At that point, with the sheet wrapped around my shoulders and my hair sticking out in short cropped patches, I looked so comical, I had to laugh. I took a lot of pic-tures on my phone. Then I shaved off what was left as close to the scalp as I could. The actual doing of it was not as sad or frightening as I had expected; I was more curious than any-thing else and weirdly detached. I didn't achieve the perfectly finished billiard ball look I had in mind because I couldn't get close enough to my scalp without cutting myself—I was left with a sort of kiwi fruit fuzz. I took more pictures of the person in the mirror. She had a nicely shaped head, I thought, but she was no one I knew.

The hair had fallen all over the sink and the bathroom floor. I gathered it up and put it in a resealable plastic bag. I couldn't throw it out. I would need it, wouldn't I, after the treatment was over? I knew that was nuts, but still I pressed the seal on the bag and placed it in one of the bathroom cabinet drawers.

I took more pictures. I took a picture of my bruised and battered breast. Then I tried on the wig and took a picture of that. It looked weird. I waited to be upset, but I wasn't really—the experience was so bizarre, so beyond anything I could have imagined, that I was more fascinated than anything else.

The hair would grow back, but in other ways, I would come out of this permanently altered, my bones and heart weakened by the drugs and five years of hormone deprivation, the shape of my body different—perhaps drastically so, depending on what surgery I had—and fatter, if the statistics on the percentage of women who gained weight during treatment were to be believed. Even the color and texture of my hair might come in differently, gray most likely, or white (like Gandalf, I liked to tell myself, in *The Lord of the Rings*). And while all of that sounded pretty scary, wasn't it just a sped-up version of the changes that getting older would inevitably bring? Would I be changed in other ways as well? Would the fear and the anxiety and the confrontation with mortality ingrain themselves so deeply that they would become a permanent part of me?

I slipped under the covers, but my head was cold and the stubble was prickly on the pillow. I got up and found the hat that Karen had knitted for me and put it on. It was soft and warm against my scalp, and wearing it I fell asleep.

Even catastrophes have routines. Gradually a latticework of daily chores provided a kind of structure to the new world in which I found myself, one in which the only thing I had to do was cope.

Henry was still a puppy then, and he took my illness hard, whimpering at the side of the bed when I wouldn't get up, or laying his head on my lap while I fought down the nausea. He forced me, more than I could myself at the beginning, not to surrender to the misery, as luxurious as that temptation was. On a more basic level, he needed to be walked, and I needed the exercise if I wanted to retain any sense of normalcy.

So we began to walk up and down the hill that led from my house to Noah Wood Road. It was only a mile each way, but it might have been Everest then. To cheer us both up, I made up a kind of chant, based on the ones the GIs used to sing during marching drills in the days when my family lived on military bases:

> *I have a dog, his name is Henery*
> *He's my friend he's not my enermy.*
> *He will come when he is called,*
> *'Specially when there's food involved.*
> *He will sit when you ask him to*
> *If he's got nothing better to do.*
> *Sound off! One, Two!*
> *Sound off! Three, Four!*
> *Sound off one two three four!*

Slowly, over the next months we walked farther, taking the right onto Noah Wood, and going up the hill, in the beginning just until the first rise, then as far as Harriet and Dean's, and finally up to where the pavement ended and the road became a Jeep trail. We stopped there—chemotherapy is cumulative in terms of its debilitating effects, and I didn't trust going out of sight of passing cars, even as sporadic as their appearance was.

In the beginning I had to stop every few steps on the uphill, and the dappled shadows, the glinting of the white quartz in the sunlight, the sound of the brook—loud as it rushed forward during the spring swell, a soft plashing in summer—distracted me whenever I felt like fainting and muffled the dreadful voices that sat in judgment day and night. They did not dare follow me on those walks up and down Noah Wood.

Because I walked so slowly, I felt, for the first time, the broad back of the earth beneath my feet, understood, if only faintly, the great consoling immensity of the world as it rocketed on its way, and it was steadying, reminding me of all that endured when much would not. I claimed the road and it claimed me, rock, stone, light, leaf, shadow, and sun. I owed it everything that was good.

Most days I saw Dean, always busy, mending a fence, unloading the pickup, feeding the horses. So, hear you've hit a spot of heavy weather, was all he said the first time I ran into him after the diagnosis, but the way he said it was comforting, containing as it did the assumption that I was up to the challenge.

Joanna, Dean and Harriet's next-door neighbor, tall and thin and unassuming, silver haired, quiet mannered, the terror of the tennis courts and a breakneck horsewoman, would come out to see how I was doing. Jill, beautiful, well coiffed, and elegant, who owned the house across the road, offered the use of her pond so I could swim without feeling self-conscious about my bald head. The young men doing roadwork or tree trimming or logging would wave a hand or indulge Henry when he reared up on his hind legs to thump his big muddy paws on their chests.

The walk pretty much did me in for the day, at least early in the three-week chemo cycle. But on the better days, I would go into the village for the endless medications I was taking, stopping at the Woodstock Pharmacy, where the staff would have cured me on the spot with their kindness. Once, when my fingernails were falling off and I could no longer get the pills of one prescription out of their fiendishly encapsulated wrappers, the women opened about a hundred of them and put them in a bottle. At the Creamery, where I sometimes went for lunch, the waitress, Tina, regaled me with tales of her annual trip to Disney World, and promised me a hot fudge sundae when my taste buds came back.

On the bad days, I stayed home and wandered through an odd sort of half-life, one in which my brain quietly packed its valise and headed for Cuba. There are many women who go through the business of breast cancer and barely break stride, who take care of young children and cranky husbands, who write briefs and run the numbers and do the laundry and remind the boss that his wedding anniversary is coming up and calmly note the dates for their chemotherapy or reconstruction or radiation sessions in their date books right alongside the eleven-year-old's soccer practice and the eight-year-old's Suzuki lesson.

I am in awe of these women, but I was not one of them.

I tried to work, to research and write the magazine pieces I had agreed to deliver and the book I was contracted to turn in, but I could not find my way to the place where I did these things. It was a little like the first months of my daughter's infancy, when, as a new mother, I felt as if all the boundaries in my world had exploded, that the safe place from which I worked and thought and sorted out emotion had vanished, my whole being invaded by the enormous fact of this baby who left only rubble in her path.

At least Zoë hadn't made me stupid. A gibbering idiot who would discuss the troubling infrequency of her child's bowel movements with total strangers, yes. But still someone who could read a book, if she had had the time, or follow the plot of a movie, if she stayed awake long enough.

With cancer it was different. In my head there was only static. When the static was really loud, I lay in bed and stared at the white birch and kept a picture of Zoë by my side, to remind me of why I was doing this. When it was quieter, more like white sound playing in the background, I knitted and listened to audio versions of Dickens novels. Dickens's characters, so cartoonishly oversize in character and personality, so heroic, so villainous, so idealistic, so energetic; the constant kinetic energy of light and dark, good and evil, tragedy and comedy, "in as regular alternation as the layers of red and white in a side of streaky bacon," as he writes in *Oliver Twist;* the druggy high of the emotional roller coaster he takes you on, the breakneck plots you follow helplessly from turn to careering turn, and almost more important, the neat, and tidy, and orderly world of his plucky and utterly unbelievable heroines prevailing over absolute chaos, evil, and bad housekeeping: Dickens saved me, gave me strength, even on days when I was so tired that it took me three hours to change the sheets on the bed.

I cruised the TV listings for anything I could stand to watch. I don't know whether it was cancer or chemo or adrenaline, but for the most part I couldn't watch TV without being acutely aware that I was watching actors acting. The veil of the imagination, the hypnotic trance that usually takes over and makes you forget the artifice, was gone—I worried about them flubbing their lines, or wondered about the props falling down, and it made me anxious for the actors.

There were two performers immune from this phenomenon—Dwayne Johnson, aka the Rock, and Vin Diesel. Luckily for me, one of their movies seemed to be playing on some channel somewhere day and night, and I watched them all in real time—sitting through the commercials made me feel weirdly safe. My favorite was *The Pacifier,* in which Mr. Diesel plays a former Navy SEAL who has to protect five lovable children from a host of bad guys. I must have watched it a dozen times and still feel a debt of gratitude to Mr. Diesel whenever I see him in a trailer or advertisement.

Most days I glued myself to the discussion boards on BreastCancer.org. The women there—scared, funny, angry, depressed, irresistible—got me through the entire process, from the forum called "Diagnosed and Scared" ("Am I going to die?" someone wrote; "I don't want to die") to the one entitled "I Want My Mojo Back" ("Hey girls, this loss of libido thing is just unacceptable. If the genders were reversed, this would be the first problem they addressed").

Over time, I noticed a shift. The fear that had seemed to emphasize my isolation in Vermont had dissipated, most of the time, permitting the return of the comfortable solitude that had set in just before the diagnosis. At least it was technically solitude—there was no one else in the house. But I had the oddest feeling that I wasn't alone.

Soon after I moved to Castle Dismal, I saw a luna moth for the first and only time. It was evening, and the creature, attracted to the light inside, attached itself to a pane of glass on one of the French doors that looked out on the woods. It was a thing of astonishing beauty—its wings were enormous, and green-blue, like a shallow sea on a sunny day, with markings in black and orange that looked like exotic eyes rimmed in kohl. The moth stayed for nearly an hour, until Henry loped into the room and startled it.

I had been lucky: luna moths, I read later, are common in North America but rarely seen because they are nocturnal and have a life span of only a few days. All summer I waited for another, though I knew such a thing wouldn't happen again. I wasn't sure I wanted it to—the singularity of the moment had been a proof that I was where I was meant to be, at a moment when such proof was sorely needed.

One night, sometime in the middle of the chemo cycle when the worst was over and I wasn't yet dreading the next round, I was sitting on the same sofa in the living room, staring out at the inky night and the woods beyond the French doors that led to nowhere. I saw something in the same pane of glass in which I'd seen the moth. It was large and pale and glowing and I was puzzled at first as to what it could be. No living thing surely, at this time of year—April nights were cold and unforgiving. The reflection of a picture on the wall? But I hadn't yet put up any pictures. I looked harder. It was an enormous white egg. No, it was a face.

An old man's face, oval like an egg, pale and nearly featureless, and bald, with a pair of thick brown spectacles delineating the eyes, a puffy face wearing a quizzical expression. My face.

I stared. Fascinated, repelled. Me and not me. All my life I had checked myself out in store windows and department store mirrors—that fascination you have as a child when you happen to see yourself, delighted, a little frightened perhaps, that the consciousness of self you have inside your head actually has a physical presence in the world, never really goes away.

That little girl holding on to someone's hand, the stringy-haired beanpole with bad posture, that massively pregnant person, that was me? Really? In my twenties there were moments I was amazed to find any reflection at all—I felt so invisible, fluid, so essentially unreal compared with other people. And then middle age, with its inevitable concessions to gravity, the time when you find yourself less inclined to seek such images out. Still, the changes come on gradually, and you make your peace with them, or at least you do until you see yourself as others see you, in a photograph perhaps. That's me? Really?

But nothing prepared me for this new version of myself, this old bespectacled egg-headed man. I called him Augustus Egg, after the Victorian-era painter who looked nothing like that reflection in the window but whose name was too perfectly suited to the image I saw in the window that night.

Augustus Egg became my cancer persona, an individual quite separate in his thoughts and temperament: me, yes, but also not me, not the woman who cried and dreamed her anxious dreams, but a doughty individual, ill-tempered, and slow moving, yes, but imbued with a Churchillian stubbornness that I would eventually come to admire.

But I wasn't very fond of Mr. Egg when he first made his

appearance; in fact, he freaked me out. I just wanted him to disappear, so I got up from the sofa and went outside to let the dog out and to look at the moon. But that night, gray clouds had partially covered it, transforming it into a torn and ragged breast, and for once I wasn't frightened or anxious or searching the ether for answers. I was furious that even the comfort of the moon had been taken away from me, that everything I cared about was gone. I was full of rage and then and there I welcomed the entrance of Mr. Egg, who didn't seem to give a damn about any of the things I worried about.

7

Last Days of the Killer Wisteria

What is a weed?
A plant whose virtues have not yet been discovered.

—RALPH WALDO EMERSON

Augustus Egg was a good companion as the spring warmed and the countryside bloomed. In the evening, the cool of the approaching night filled the hollow with the scent of lilacs and issued such an imperative to joy that I wanted to run into the house and shut it all out. It had nothing to do with me, with the grayed-out landscape of illness. But Mr. Egg just laughed— you have to stand here and take it, he said, stand here, among the blue-black shadows of the blossoming trees, drinking in this scent and this night, and understand that the world in its infinite wisdom will go on, spilling its prodigal beauty and unmerited pain and what a very good thing that is.

As the weather began to settle, the workers waiting for their sandwiches at George's butcher shop were beginning to talk about barbecue weather and the first of the weekend people were back to order crown roasts and legs of lamb. On Pleasant Street, in the sunshine splintered by the chill spring wind, a small pod of teenagers walked, dripping wet and barefoot, rushing the season with a dip in the Ottauquechee. At

Runamuck, Cathy's barnyard was in riot, with turkey chicks poking their ugly little heads out of eggs, ducklings strolling about, geese squawking, baby goats leaping over tricycles and wandering into the house, children careering about on bicycles with training wheels. On the green slope above, a proud red vixen sat, showing off three cubs nursing at her side, while improbably, on just the other side of the fence, a flock of sheep grazed. It seemed as if everyone in the village was outdoors, dressed in sun hats and mud boots, crouching over flower beds, wielding trowels and hoes and hoisting large bags of compost and fertilizer.

I stayed inside, reading about life expectancy rates and treatment options. On good days, when I was in Marcus Aurelian mode, I was simply grateful to have a shot at surviving. On others, I thought about the physical changes that surgery would bring, about whether or not to have breast reconstruction—was a fake breast a surrender or a reassuring return to normalcy? I still didn't know whether I would even need a mastectomy—the tumor showed signs of diminishing—but I couldn't stop thinking about it, because it was a way of thinking about so much more.

The idea of mastectomy without reconstruction was savage and drastic, and defiant—it was a jump off a cliff. I liked the extremity of it. Besides, there was something egregious about a fake breast, if only because I resented the idea that you need to have it—or any—to be "a whole woman," as the preconstruction literature stated. It also seemed vaguely retrograde, as if I was trying to hold on to something I no longer really needed. Mastectomy appealed because it seemed, at least to me, a perfect and irrefutable signal of the end of sex, a definitive answer to the nagging questions of love and loneliness, to my frustration that I could not yet give up on the idea of a shared life. My doctors tended to just roll their eyes when I ranted on like this, figuring, correctly, that my attitude would change when the time came.

Maybe it was this obsession with surgery, maybe it was a reaction to all the life springing up around me, but I developed a visceral need to cut things down. Anything; really, it didn't much matter what. The last of the stubborn pockets of snow were finally receding from around the house and the earth was giving way to the advance of the new season. Some days it seemed as if everything was falling down: a tree limb broke off high above my head and fell a few feet from where I was walking; the posts holding up the woodshed lost their hold in the wet earth and collapsed, sending some of the neatly stacked logs rolling down the hill to the creek, to the great delight of the beavers.

I wanted my part in this gleeful destruction, and on days when I had the energy I would put on a pair of muck boots and throw an anorak over my nightgown, and machete in hand, Augustus and I would go forth and do battle.

Most of the time I went out back to where the tall thorny raspberry canes tore at my clothes every time I hung the laundry, and obscured the view of the creek below. I whacked away blindly, stopping only when I could no longer lift the blade. But the real target lay in the front yard.

From the moment I had moved into Castle Dismal, I hated the ugly thick-stemmed green vine growing on the right side of the house near the porch. Over the past five years, it had grown out of control, *Little Shop of Horrors*–style, coiling up the

porch columns. For the most part I left it alone, unable or un-
willing to challenge its progress. But there were limits to my
passivity. Whenever it extended its grasp to the porch furniture
I would take action, snipping away the tendrils that were coiling around the rocking chair in which I liked to sit and hatch
dark plots for its eventual demise.

I had assumed for a long time it was simply a great ugly
weed, but recently I had been informed that it was in fact
a wisteria vine, and that bit of news had stopped me in my
tracks. A wisteria conjured an entirely different set of images
than my rampaging weed: it bespoke graciousness, quaint
well-tended cottages and the blooming, cultivated lives within.
I would have to change my attitude, I decided, not without regret, for I had enjoyed hating the plant as much as I did. Now
I would have to tend my wisteria and encourage it to bloom
for me. Its delicate blue flowers would soften the severely practical lines of my house. It would thrive under my care, and I
would learn to love it, because wasn't that why I had come to
Vermont? To learn to take care of things, to be the kind of person who tended a garden with patience and care, and weeded
and pruned and then basked in the results of her stewardship,
proud that she had created this little Eden on earth?

No, said Augustus Egg. It's ugly and you hate it.

No, it's not, and no, I don't, I said. At least I won't when it
flowers.

It won't flower, said Augustus.

It won't flower because it hates me, I said, and a tear trick-
led down—whether from exhaustion or from the Taxotere that
kept my left eye watering constantly was unclear.

It won't flower because it's ugly and horrible and keeping

it around isn't going to make you into the kind of person who keeps a beautiful garden or recites poetry by heart or opens her house to the public on the gracious home tour and it isn't even going to turn you into the person who wants to be that person, said Augustus.

You're right, I said, and got the machete.

I whacked away at the wisteria for a couple of hours. The machete's silvery sheen and keen edge, which I had so carefully honed in the beginning, were duller now, but the gleam of the metal against the matte green of the thick stems was still cruel and beautiful to look at, and the increased resistance of the stalks and the heaviness of the handle and the immediate ache in my arms turned each swing into a howl of anger and rebellion and joy and pleasure. It started to rain and still I whacked away. My hands were bloody from the thorns that had grown up around the vine and my feet were muddy and Henry danced in and out, trying to help, growling and pulling at whatever piece he could get ahold of. The wisteria grew on a trellis that had originally been a white fan-shaped structure, but which had weathered to a shabby gray and buckled under the weight of the monstrous vine, and so I got out the hatchet and hacked down the trellis as well. At the end the trellis was an ugly splintered mess and the vine was a stumpy ruin of its former self and everywhere was mayhem. The tangled vines I'd torn away from the porch railing lay in great heaps, threatening to suffocate the budding daylilies. I saved the daylilies but left the rest where it lay, and went into the house content.

Catastrophes provide a pair of parentheses in which to live apart from real life, depositing you rather abruptly on the sidelines for a bit while normal life continues to eddy downstream. And, like Dr. Johnson's hangman, it does serve to concentrate the mind wonderfully. I thought a lot about the disorientation that had brought me to this peculiar place while I drifted in the chemical fog and set the rhythm of my days to the timing of the next infusion, the rounds of white counts and blood tests, the progress of the tumor toward extinction.

Eventually the treatment would be over, and whatever the long-term outcome, I would return to real life. The current crisis forced a necessary optimism, a cheery determinism to get through this mess with as much courage as possible and back to life as it used to be. But did I want that life back? A brain beset by chemicals may not be the sunniest plateau from which to review the past—the bleak, scoured emptiness in which I lived shone an uncompromising light on what had come before, unmediated by softer memories. But I welcomed the harshness.

Looking back, I could see only failure and loss, and heard only the voice that asked, with increasing insistence, that age-old midlife question: Is the best of it over? The chemo had made me look my age (or, as Dr. Ruden had put it when he saw me during that time: You look great! All that weight you've put on has gotten rid of the wrinkles on your face!). And while the chemo may have paralyzed my brain, how long had it honestly been since I had taken any real interest in my work before I got sick? What was I living for, apart from blind instinct, apart from Zoë? I wasn't thinking about these things in anguished, apocalyptic terms, but in the cool dry logic that Augustus Egg favored—to him it merely seemed like an interesting question. He asked a lot of them: How do you judge if your life is a failure? I had often told myself mine was, but how did one define failure? Is it only about whether you have hit all the marks, the ones the others hit, and when did you decide that they were your marks as well?

When I was young and insufferable, I wrote a fair number of newspaper stories about cultural phenomena who had faded from the headlines but still had the temerity to ply their craft, what one editor called the surprise-he's-not-dead profiles. I liked working on these stories; the has-beens were invariably more interesting than the ones who were still floating along like those Japanese puffer fish, engorged on the attention and oblivious to the notion that they could lose it all. But I always felt a little sorry for the subjects who insisted they were happy with the way life had turned out, that they had found a richness or a source of satisfaction that more than made up for what they had failed to achieve or thrown away due to their own weakness, or excess, or bad luck. I didn't believe them, that's all.

But now I wondered at my youthful condescension (and the fear that fueled it): Wasn't it just a reflex, an echo of the sort of assumptions and received wisdom that we accept until we decide for ourselves which of them are weeds and which wisteria? I wasn't sure then that I had ever made that choice. Or had it always been the expectations of others that drove me forward, the desire to please, the shame when I failed to do so? And did being a failure— if that was what I was—make my life any less valuable? Less pleasurable, for that matter?

The questions never stopped; they followed me down the highway in the car, they kept me up at night in the dark dead silence, they cackled at me from the rafters. One afternoon while driving down my favorite road, Church Hill, the voices were more brutal than I could bear, drowning out even Mr. Egg, so loud, so merciless, that I had to pull over. I tried to do a meditation I had learned from *The Places That Scare You*, by the American-born Buddhist nun Pema Chödrön, but which is common, I believe, to most Buddhist philosophy. It is a simple exercise in compassion, wishing that all who suffer may find peace. When breast cancer scared me, the act of wishing that all the women who were as scared as I was would feel better always calmed me down. So I tried it there, on Church Hill Road, parked next to a tumbledown barn looking over the green frosted hills that were just coming alive with bright yellow wildflowers. May all those who feel as lost as I do find their way, I said. Yes, said the voices, all of them but you.

As I started the car I wondered if it was possible to truly forgive yourself for your sins, real or imagined, whether I would ever escape the regret in which I steeped my version of the past. I knew by then that I had not left New York because I was ready for something new. No: I had run away from my old life out of a shame so large I hadn't even seen it for what it was.

All things fall and are built again . . .

And that, I finally realized as I headed toward home, was the giddy liberating glory of it: all things fall, even the hand that holds the scourge, done in by the incontrovertible insistence of the present. I drove on in the cool blue shadow of the hills, into the gathering dusk of the evening, as the diamond points of the first of the night's stars pricked the sky. Maybe they were all true, the crimes and shortcomings of which I convicted myself, and maybe, just maybe, it didn't matter, because at that moment, when I crested the last hill and turned into the driveway and stepped out of the car into the lilac-scented night, I was nearly knocked over by the beauty of a world in which my part mattered not one whit. All I could feel then was a surge of happiness and gratitude; I was that glad to be alive.

Go figure.

Cancer is a good teacher. It forces you to understand what you should have known when you were healthy: there is little time left and none at all for regret. I might not know what I wanted but I was beginning to know what I didn't. There were a lot of vines to be uprooted, most of them snaking through places in which a machete would be of no use. But I suspected that underneath them might lie a path to equanimity.

The rage began, as cataclysms often do, suddenly, like the first drops of rain that touch your cheek as you walk down a busy street. You blink, and look to the sky, and there is not even time to register the idea that a storm is coming before you are drenched to the bone.

For me, the moment came one morning in late October as I lay alone in a large dimly lit room, on the other side of heavy double doors emblazoned in yellow: DANGER! RADIATION. I was naked to the waist. My feet were strapped together with a large rubber band, my arms raised above my head, slightly bent at the elbow, my hands in stirrups just above my head. On the ceiling, backlit by fluorescent lights, was a lurid photograph of an island cove at sunset, mountains rising majestically in the background, all of it bathed in a kind of acid purple. The sound system came alive with some bouncy, banal 1970s song—it probably wasn't "Afternoon Delight," the irony would have been too perfect, but it might as well have been.

It was my first of thirty-seven radiation treatments.

A few moments earlier, I had walked, more nervous than
I knew, into the dark room, which was nearly empty except
for a high, padded table and an enormous white machine that
loomed over it, consisting, from my limited perspective, of a
giant arm, at the end of which was a large lens that opened and
shut at regular intervals, like a slowly blinking black eye. It was
made of a shiny dead white plastic and looked evil and men-
acing, like a weapon that might defend Darth Vader's storm
troopers in *Star Wars*.

I had climbed up onto the table and lain down, tense and
expectant, until one of the technicians reminded me that I had
forgotten to remove the pink wraparound hospital gown I had
been given to wear over my jeans. "You forgot to take off your
johnny," she said.

That was when I had felt the first tickle of irritation that
was only the fanfare for what was to follow. It was that word—
johnny. It was such a stupidly cute word to designate the thing
I wore, the thing that was an emblem of what was happening to me, one that erased everything distinctive and individual about the women who wore it, the increasing number of
women, who waited their turn each day out in the pleasant
waiting room strewn with sofas and easy chairs and small tables covered with magazines and half-done jigsaw puzzles, the
women who were sitting there now, their bald heads hidden
by turbans and baseball caps, and wool hats and scarves, the
women with complexions grayed by chemotherapy and insurance worries and fatigue and pain. Women sentenced to wonder for the rest of their lives if what was happening to them was
for the only time or only the first time. It wasn't a "johnny" we
were wearing. It was the uniform of the prisoner.

What was wrong with me? I wondered, puzzled by my reaction to such a dumb little detail. I was not the shell-shocked neophyte of eight months earlier, reeling from the diagnosis. I was a veteran, of eighteen weeks of chemotherapy, of surgery and recovery, of countless waiting rooms and blood tests and bone scans. I was accustomed to hospital procedures and the emotional ups and downs. I thought I had learned to take it all in stride.

Besides, everyone, even the doctors, had said radiation was a cakewalk compared with the other treatments. "We like to think of radiation as chemo's kinder, gentler cousin," a radiologist had told me. Veterans of breast cancer treatment agreed. Radiation made you tired by the end and it was a great gobbler of time, five days a week for seven and a half weeks, especially if you lived forty-five minutes away from the hospital, as I did. But that was it.

And so I had breezed through the initial, pretreatment preparatory session, when the technicians had mapped and measured the tumor site and then tattooed my chest with tiny permanent blue freckles that would serve as markers for the radiation beams. I joked about being transformed into a medieval map, the ones where the dangerous and unknown places were marked with the warning HIC SVNT DRACONES, this way there be monsters. And cancer was a monster, against which we had only these starkly primitive options: to burn, to poison, to slice away. With that thought in mind, radiation had reduced itself in my imagination to a mere inconvenience, one that entailed daily visits to the hospital, a minor sunburn, and possibly fatigue. But nothing like the nightmare of chemo. A nuisance, yes, but nothing more.

The three radiation technicians spent a great deal of time getting me into exactly the right position to ensure the accuracy of the rays, shifting me about on a sheet while I lay passively on the table, because misdirected radiation could permanently damage heart and lungs. Finally they were satisfied with the position—they would take some X-rays, they said, and then administer the radiation, which in itself would last only a few minutes. They left the room, closing the door behind them.

A whirring sound and then the enormous white arm swiveled around, its great black eye trained on my left breast. I shut my eyes tight and tried to breathe, but all I could think about was the position I was in and why it seemed so familiar. Of course: it was the position of erotic welcome, of a woman lying in bed, smiling at her lover. A wave of humiliation surged through me. I lost all perspective; it didn't matter anymore that I was lying on this table in this room of my own free will, and to receive a potentially life-saving treatment, that I was lucky to have such an option, that I had made a choice. All I knew in the place where I lived that has nothing to do with logic was that I was naked and exposed, that my privacy, the memories of the pleasure my body had both given and received, had been violated, mocked, and debased. The fact that I was not forcibly restrained, that I could have walked out of there at any point, only made it worse. I was a willing participant in my own humiliation.

Dissonant images clashed in my head—Goya's naked *maja*, the black-and-white photographs of French women who had slept with Nazis during the war and who, at liberation, had their heads shaved and their clothes stripped off, and were branded with the word *collaboratrice.* I thought of my own bald head and I fell apart.

My rib cage heaved and the tears ran down. I tried to remain still, but it was no use—finally the door opened, admitting the three now worried-looking young women and one very brisk doctor, who told me, with all the compassion of a codfish, to sit up and get ahold of myself, because I was ruining the X-rays. It worked, sort of. I hated her instantly, and hate is a much colder emotion. I lay back down, every muscle turned to marble. The technicians and the doctor left the room. I shut my eyes tight against the lurid violet photograph and the cold black eye that moved and whirred and blinked and finally it was over.

I stumbled out of the room in a blaze of tears, having shaken off the abashed young woman who had tried to help me off the table. Then anger temporarily gave way to mortification. I had always prided myself on my stoicism, at least in public, and now I was scaring twenty-year-olds. I couldn't look anyone in the eye.

During the drive back to Castle Dismal, there was a small, still-sane part of my brain that watched in wonder as I screamed myself hoarse, consumed by a rage I'd never felt before. It was October and Woodstock was filled with elderly tourists out to see the foliage. Normally the leaf peepers struck me as rather sweet and a little sad, dressed as they were—as we all seemed to be now—in their play clothes, old shanks protruding from baggy shorts, sagging breasts in shapeless T-shirts. That day I wanted to shoot them all.

Back at the house I went straight to the breast cancer discussion boards on the Web, and found some comfort there:

other women reported reacting as I did, and for a little while
I was tremendously relieved—I wasn't a lunatic, or at least not
the only one. Then I got angry all over again and nothing—
not my impressive collection of psychological and Buddhist
self-help books, not the Chopin Nocturnes, not even the Rock
himself—could make a dent in my rage until the double dose
of Xanax kicked in.

The next day was slightly better. There was an older techni-
cian that day, a woman about my age, whose compassionate
smile had a soothing effect. She told me that I didn't need to
wear the "johnny" if it bothered me; I could simply remove
my top on the table. I got through the session without crying.
But as soon as I left the room the anger hit again, and I was
blind with it. I sat in the parking lot, the special one for cancer
patients, and did a crossword puzzle until I was calm enough
to drive.

In retrospect, I can see that my anger was simply one ten-
tacle of an overwhelming emotional fatigue that was almost
inevitable after eight months of treatment. I had let myself
think that the worst part of having cancer was over, and had
felt the first quickening of a relief, which of course was premature. I was like a marathon runner who thinks she sees
the finish line, only to find out she has five miles more to go. I
had nothing left. It makes sense now, but back then my anger
terrified me: it came from nowhere and I couldn't seem to
control it. I needed help.

A few days later, I knocked on the door of a neat white
frame house that sheltered under the hill behind it with the
proprietary air of a cat occupying its rightful place in your fa-

vorite armchair. I was met at the door by a small, softly smiling woman who was to be my Reiki therapist. Joanne's white hair was tucked back in a bun and her blue eyes reflected a tranquility that could have calmed a cobra.

Joanne led me into a dimly lit room with the low ceiling and close proportions characteristic of very old New England houses. I took off my shoes, climbed onto a high narrow bed, and was soon cocooned under a heavy handmade blanket. There was no sound except for the creaking of old wood and the occasional scrabble of small creatures looking for shelter from the cold.

Reiki is supposed to work even if you're asleep. Advanced practitioners can supposedly do it over the phone. I closed my eyes and tried to meditate while Joanne gently placed her hands on my eyes, ears, and head, working her way down to my toes.

The session lasted ninety minutes. At first I felt only the warmth of Joanne's hands, and the impending hyperventilation I always feel when I try to concentrate on my breathing as one is meant to in meditation. But as I relaxed, I sank deeper into a state that was somewhere between sleeping and waking, what scientists studying touch healing call a liminal state, which is said to resemble spiritual trances or self-hypnosis.

Gradually all the emotion that had been roiling around and through me came to Technicolor life, assuming shapes that twisted and transformed themselves. The rage was a roaring, pounding wave, slamming into a rocky jagged beach. The anxiety was a rain of needles falling all around me. Then the scene changed. I was inside a dark warm cave, but glaring at me from the darkness was a weasel with sharp, bared teeth, poised to attack. I was afraid of the weasel, and my mind cart-wheeled away from the cave through a jumble of other images until I finally felt safe enough to return. The weasel, much to my astonishment, was asleep. I wanted to touch him as he lay there, curled up, sleek and soft, but I knew that if I did, he would return to his saw-toothed snarling ways. Still, he did look pretty peaceful.

In my strange dream state, I kept looking for a place to be safe, while at the same time being vaguely aware of Joanne's hands placed on my stomach, and then around my knees. Eventually I left the cave and was standing on a hill looking into the woods that covered the slope beyond. There was the weasel, running in great looping circles. He was free, I thought, and I was thrilled. And yet, to my surprise, I was also sorry to see him go. He looked so alive. That's when I noticed that in my hand I held a long leash that kept us connected. I under-stood what it meant. The rage I had felt was scary, but it was also powerful; it had given me an energy I hadn't felt since the diagnosis. A part of me didn't want to let that go.

Joanne ended the session by cupping my heels and then my toes in her hands, which left me with a sensation of tremendous safety. Finally I opened my eyes and she handed me a glass of water. She asked me how I felt. I thought about it. I wasn't exactly a cup of softened butter, but I was much more relaxed. Having been given shape and form, my emotions no longer consumed me. I knew the anger was still there somewhere, but it was contained now. It wasn't me.

The next day's radiation session was easier. I was still anx-ious, and after three different people in the reception area told me to go to my "happy place," I was cranky, but calm. On the table I closed my eyes, and this time I saw myself in a room that was not unfamiliar. Then I recognized it: it was the room in Castle Dismal I had planned to make into my office. In reality the room was filled with unopened boxes and bags, but just then I saw it as it could be—neat, orderly, a place to work and be whole. A promise of life after cancer.

In 1527, Álvar Núñez Cabeza de Vaca, a Spanish nobleman, set sail for the New World, one man among a grand armada of men intent on conquest and glory. He served as one of the king's treasury officials accompanying the conquistador Pánfilo de Narváez on an expedition bound for the northern rim of the Gulf of Mexico and the war with the indigenous population that would follow in its wake.

As they were approaching landfall, a great storm wrecked the fleet. What had begun as a well-supplied expedition of six hundred men and ten women eventually dwindled down to an epic struggle for survival by four men over the course of nine years in a vast and violent land.

Bad luck and their own folly plagued them from the moment they landed. Because the expedition's navigators didn't know where exactly the remains of the fleet had landed, de Vaca advocated they hug the coast and keep to a course that would establish their bearings. But Narváez was eager for gold and sent his men in the direction that rumor said it could be had. Both the land and sea forces marched into the immense Florida swamp, a nightmare trek that left them decimated by disease and their battles with inhabitants less than eager to be occupied. By the time the expedition reached Apalachee Bay, there were only 242 men. Hope, however, had survived.

Starving, wounded, sick, and lost, they decided to take to the sea. They slaughtered and ate their horses—an act of desperation that stole from them their very sense of identity, of what they had been born to be—caballeros, knights, set apart and above. They fashioned a bellows from deerskin, melted their stirrups and horseshoes, their spurs and shields, and turned them into nails and tools. With these they constructed five rough-hewn boats and set sail for what they thought was the coast of Mexico and the other Spanish forces they believed to be nearby. In fact, they were 1,500 miles away.

Cabeza de Vaca commanded one of the little boats, each of which held about fifty men. They ran through what supplies they had all too quickly as they followed the coast westward to the mouth of the Mississippi River. Then came the final blow to whatever remained of their original plan: a strong current swept the boats away from land and into the teeth of a hurricane. Only two of the boats survived.

The surviving two, with about forty men, including de Vaca, came to grief near what was probably Galveston Island. The former conquistadors had other names for it: they called it Malhado—Misfortune; they called it The Island of Doom. There they tried to repair what was left of the boats, using their clothes to plug the holes, but fate wasn't done with them—not by a long shot. A large wave crashed on the shore, sweeping the boats away as if they were bathtub toys.

The little band of survivors became smaller still. Tribes living along the coast enslaved the few who were left, tribes whose names can only be guessed now, extinguished as they were by the subsequent waves of men who came to take their gold and save their souls.

At some point, de Vaca and his companions escaped from their captors and set out for . . . what? They had no way of knowing, save for the fact that there were other Spaniards out there somewhere. They were captured again, and escaped again. Soon there were only four men left: de Vaca; two other Spaniards, Andres Dorantes de Carranza and Alonzo del Castillo Maldonado; and Estevanico, a luckless Berber who had been taken in slavery from his native Morocco (and who would survive this adventure only to die at the hands of natives as a member of an expedition led by Francisco Vásquez de Coronado, another Spanish conquistador).

De Vaca and his companions were as lost as any humans have ever been, naked and defenseless in an unknown country in a harsh climate, without a map, without a bearing, without a clue. They had no words for most of the animals and foliage and wonders they saw; they were the first of their kind to see them.

They stayed lost for nine years. But somewhere along the way they shed the narrative of capture and escape; transformed by hardship, time, the wild and brilliant landscape, and whatever bargains they had struck with their God, they were no longer helpless witnesses to their own catastrophe.

Cabeza and his men traveled hundreds and hundreds of miles barefoot and naked. They explored, as nearly as anyone can piece together, much of Texas, parts of New Mexico and Arizona, as well as what are now the states of Tamaulipas, Nuevo León, and Coahuila, in northeastern Mexico.

From there they walked south, through Coahuila and Nueva Vizcaya, down the coast of the Gulf of California to Sinaloa. They became traders for a time, traveling from tribe to tribe, learning the languages, bartering goods from one place to another, forging a kind of life. Eventually, after shedding his sunburned skin a dozen times over, Cabeza de Vaca became "a child of the sun," in the language of the people there, a holy man, a seer. He healed the sick, he saved the crops, he raised the dead. The people gave him copper rattles, coral beads, turquoise, arrowheads, and, once, six hundred deer hearts. He and his companions were no longer alone: they were followed on their journey by crowds of the faithful who believed de Vaca had the power to heal and destroy them. Which, of course, was exactly what the lost men had themselves believed when they set out from Castile that fine day so long ago.

Reading his narrative, which he would write and rewrite the rest of his life, you wonder if de Vaca ever lost the man he once had been, if the lust for gold and glory that had propelled him across the ocean ever burned off like a morning mist in the heat of the day.

The facts would suggest otherwise.

Eventually the little band arrived in a place he called "the Village of Hearts." They were in New Mexico at that point, a terrifying, empty place, deserted, he writes, by those who lived there because of recent depredations by vicious men on horseback. De Vaca knew he had found his people.

He left his followers, promising them he would stop the killing, the slavery, and the theft of their land. The next day he and the Berber Estevanico and eleven of the natives walked thirty miles until they met up with a company of Spanish soldiers out hunting slaves.

At first, the Spaniards couldn't believe de Vaca was one of them and promptly tried to enslave his companions. De Vaca in turn was so angry that he and his fellow travelers took off, leaving behind pretty much everything they had brought with them.

The Indians, for their part, were mystified. They simply couldn't believe the men on horseback belonged to the same tribe as the naked god with whom they walked. As de Vaca explains, "we came from the sunrise, they came from the sunset; we healed the sick, they killed the healthy; we came naked and barefoot, they clothed horsed and lanced; we coveted nothing but gave whatever we were given, while they robbed whomever they found and bestowed nothing on anyone." For de Vaca it must have been like walking through the looking glass—what these men were, he had once been.

And, to a degree only he could know, still was. Although in the beginning he could not bear to sleep on a bed or wear European clothes, de Vaca sailed back to Spain in 1537, ten years after he left it.

Perhaps the most astounding aspect of de Vaca's life, and the most difficult to understand, from an idealistic—or naive— point of view, is his return to the world he once inhabited and to his position within its elaborate codes of behavior and measures of success. He had gone through an extraordinary transformation—a redemption, in the eyes of those who see him through the lens of history's bloody coda—and yet he went back to the court of King Charles I, and took up the ways of his old life. There he wrote the account of his travels, having taken care, he said, to remember everything faithfully, so that, if God so ordained, "I would be able to bear witness to my will and serve Your Majesty, inasmuch as the account of it all is, in my opinion, information not trivial for those who in your name might go to conquer these lands . . . and bring them to knowledge of the true faith and the true Lord and service to Your Majesty."

Eventually de Vaca returned to the New World, as governor of the Province of Rio de Plata in what is now Paraguay. His tenure ended badly—he was arrested, imprisoned, and sent back in chains to Spain, where he was tried on charges of official misconduct, of mistreating the Indians and of raising his own heraldic standard instead of the king's, though some accounts insist that his true crime had been his attempts to protect the indigenous people in his charge.

Cabeza de Vaca ended his days back in his home in Jerez de la Frontera, Andalusia, old and penniless and brokenhearted, say the more apocryphal accounts, busy ransoming his nephew from the king of Algiers, according to others.

Cancer veterans warn you that the year after treatment is, in some ways, worse than the year in which you are actually fighting the disease, and there is something to that. While you are in treatment, you have a job—there are appointments and pills and side effects, there are blood tests and bone density tests and tests of character and simple endurance, decisions to make, small and large.

I was lucky—the cancer had retreated enough to enable me to avoid a mastectomy and there was no evidence of the disease in the lymph nodes. But still, when it was over, I was caught up short, and the first few months afterward were spent in a kind of free fall. Augustus Egg had made a quiet departure without so much as a good-bye, the adrenaline on which I had lived subsided, and all the emotions that it had bullied out of the way returned. Oh my God, I thought to myself, curled into as small a space on the sofa as I could manage, I had cancer! Everything seemed tentative then, as if I were walking on very thin ice, as if a splinter of light could pierce me. For a while, I could not even boil an egg—the commotion of the water, the violence of the act, the injury to what lay within was too unsettling. How odd it sounds now. But that is the way it was.

I went into hedgehog mode. I had a family of them living with me. My first year in Vermont, there had only been one, a male, I think, who, in the warm early-autumn evenings, would noddle around the rocks and weeds in the space outside the back door where a patio might be someday, if I ever became the sort of person who had a patio. He would look up when he saw me, acknowledging my existence, but rather like the mice in the pantry—or the men who plowed the road in winter, for that matter—he didn't appear to take too much notice.

But during the spring, while I was sick, he must have decided to settle down, because now he had a mate in tow and a fine-looking family of little hedgehogs who had taken up residence in the woodshed that abutted the garage. They were renovating: a small chink in the wall between the two was growing larger every day, and when I would try to impede their progress by moving a jug of distilled water or a log in front of the opening, the evening would resound with the rhythmic thumping of hedgehog tails battering down the latest obstacle to their ambitions.

Harriet's husband, Dean, advised me to get rid of my new tenants immediately. He was the gentlest man I had ever known, but though he could talk for hours about the birds he observed at his feeders in the spring, he took a less patient line toward the more destructive members of the animal kingdom.

While some locals used hedgehogs for target practice, Dean took a more pacifist approach. He trapped his own beasties in Havahart traps and drove them to the woods near Reading, the next town over, where he let them go. First mom, then dad, and finally the offspring. When I worried about the pups or kids, or whatever you call hedgehog babies fending for themselves, Dean shrugged. "They're alive, aren't they?" he said. "Best I can do."

He was right, of course, but I let them be, beyond reading up a little on their habits. Hedgehogs, when they are threatened, curl up into a ball, their soft and vulnerable bellies disappearing deep within a sphere of quills that offer a mouthful of pain to any predator. In those first few months after treatment I could think of no better defense against the frightful option of normal life. I didn't go out, or answer the phone, or listen to music with words—I wanted no emotions or moods to intrude, no opinions or promises or plans.

But life, whether you like it or not, rebuilds itself. You begin with scaffolding; the daily routine—dishes washed, bed made, a numb trip to the supermarket. Then one day in a parking lot a chance encounter with a friend yields a dinner invitation you are not permitted to decline. You drive home from an evening spent around a kitchen table with friends laughing in low voices, and for the first time you notice the clouds scudding across a nascent moon. Something quickens. And gradually you find yourself inhabiting each day more fully, savoring the lift of leaves in an autumn wind, two dogs sleeping in a patch of sunlight, the graceful arc of a young man leaping from a pickup truck. And you promise yourself you will take nothing for granted ever again.

Catastrophe can change you but it doesn't turn you into a better person, at least not for long. After the end of treatment, after the effects of the drugs and radiation have begun to wear off, after the fatigue and the depression begin to lift, you find yourself sitting in a traffic jam on the highway cursing the traffic, impatient, eager to get moving. You remember how, just a few months before, you would have given anything for the normal routine of a life in which a traffic jam is the worst thing that will happen to you that day. You had sworn when the chemo beetles were eating away at your brain, your mood, your sense of self, that you would never take the minor inconveniences of life seriously. But you do. And that in fact is how you know you are moving beyond the shadow that cancer or any of life's big curveballs cast: you have regained your inalienable right to be grumpy about the absolutely unimportant.

But still—in those first few months after treatment, I thought often about Cabeza de Vaca, after he returned from his wanderings and resumed his place at the Castilian court. I imagined him buckling on his ceremonial armor, before presenting a petition to the king. Perhaps he took one last look in the mirror just as the sun escaped a cloud and bounced smartly off of the polished metal of his breastplate, blinding him, reminding him for an instant of the infernal glare of the sun in that savage land that had peeled away his skin, and laid bare his unarmored soul. Did he miss the naked holy man, the disaster that had loosed him, had taken away his compass and freed him from the dictates of so elaborate a life?

8

A Sense of Direction

It is better to think of a return to civilization not as an end to hardship and a
haven from ill, but as a close to an adventurous and pleasant life.

—SIR FRANCIS GALTON, *The Art of Rough Travel*

O n the other side of Boreas, the North Wind, there was
once a hidden paradise, a land temperate and fertile
and beautiful. The people who lived in Hyperborea were
blessed, and their lives were devoted to Apollo, whose round
marble temple stood at the center of a sacred grove of cedar
trees. "The Muse is not absent from their customs," wrote the
Greek poet Pindar. "All round swirl the dances of girls, the
lyre's loud chords, and the cries of flutes. They wreathe their
hair with golden laurel branches and revel joyfully. No sickness
or ruinous old age is mixed into that sacred race; without toil
or battles they live without fear of Nemesis."

No traveler ever returned from this paradise. To enter Hyperborea, one had to endure bitter cold and privation, and to
pass through places of dread and horror—Ierne, for one, or as
we know it now, Ireland, whose inhabitants "consider it honor-
able to eat their dead fathers and to openly have intercourse,
not only with unrelated women, but with their mothers and
sisters as well." Past Ierne lay Ultima Thule, the most distant
place on earth, a nightmare land, where, some said, one could
find the gates of hell.

Despite, or more likely because of, the dearth of eyewitness accounts, the belief in this perfect paradise has proved so beguiling that it has echoed through the centuries. In Mary Shelley's novel *Frankenstein; or, The Modern Prometheus*, the eponymous hero describes the Hyperborean paradise to his sister: "I try in vain to be persuaded that the pole is the seat of frost and desolation; it ever presents itself to me as the region of beauty and delight . . . there snows and frost are banished; and, sailing over a calm sea, we may be wafted to a land surpassing in wonders and in beauty every region hitherto discovered on the habitable globe." The location itself has proved flexible: when Admiral Richard Byrd flew over the South Pole in 1947, rumors abounded that he had caught a glimpse of a mysterious subtropical land.

There is something stubbornly human about this longing for a paradise on earth (and our need to make a hell out of other, more accessible places), as if our troubles were a product of geography, as if we could escape them if only we left them far behind.

On a warm Thursday afternoon in March, a year to the day after I had come for my first session, I left the office on the seventh floor of 30 East Sixtieth Street with a big bouquet of yellow roses. A surprising and touching gesture, by which the three women who worked there—Dr. Reichman; Jane Brown, the oncological nurse; and Ana Oliva, the unflappable office manager—marked the ceremonial end of treatment for each of their patients.

I walked out a little dazed. I don't think it had occurred to me that I would ever stop coming to New York every three weeks to have a needle placed in my left arm. After the humiliation and fury of radiation had subsided, after the depression began to lift and food no longer tasted like weed killer, after the first pale shaft of hair had given way to a half inch or so of curly softness, the trips to Reichman's office were no longer filled with the dread of the sickness that was to come. Instead the office had become a safe harbor, a place where everyone understood and no one had to explain. I realized, as I sheltered my roses from the crowds on Madison Avenue, that I would miss the place.

It was a beautiful day, made giddy by the brilliant cacophonous sunshine—light has a clamorous quality in the city, bouncing off windshields and storefronts and traffic lights, shouting for attention. On impulse, I decided to walk the thirty blocks to the garage where I'd stashed the Jeep, cradling the roses like a piano recitalist who had made it through a particularly intricate bit of Schubert, taking in the city in a way less fettered by memory and experience than it had been in a long time.

Then I headed north toward—home?

As the Vermont hills hove into view, a light snow had begun, and the heavy gray sky subdued any lingering sense of ceremony. In my other trips north I had always welcomed the sight of those hills, but now they stirred more ambivalent reactions.

When I had first started coming to Vermont from New York I loved that moment on Interstate 91 where Massachusetts yielded to Vermont and you knew without looking at any highway sign that you were in a very different place. It was in the light and the air and the blurred outlines of the hills on the horizon, and something in me always lightened. But not this time.

Like any of life's great storms, cancer blows a hole through your life. In the aftermath, you pick your way through the debris, salvaging what you can, leaving what you no longer have the room or energy to take. For a short time, you see things remarkably clearly, no longer quite so blinded by your own mythology or the stories you have concocted to make sense of the life you have led.

What I was beginning to see now was that Vermont was not the place where I could elude the gravitational pull of lifelong worries and regrets, a paradise of simple living. Those starry visions had died quick enough and I'd ended up in a swamp of my own making. Cancer eliminated that swamp of grief and loss and animated gophers, for which I was very grateful. More than that, it had given me something I hadn't bargained on: it gave me back myself. Not the one I had moved to Vermont to become, the recluse who needed no one, nor the one I left in New York, that wretched failure, but someone else both familiar and unfamiliar, gawky and barely recognizable—the self that lies beneath the roles we all accrete as we go along, layer upon layer—good girl, bad girl rebel, failure, success, and exile, all of them. The self I had fought so hard to save was none of these things, not the writer or the mother or the woman who didn't know what mulch was. It was a being that simply wanted to continue to be. That was it.

That self—an identity so bedrock it didn't even have a gender—would disappear again, the moss and lichen of illusion and delusion, regret and longing, would grow over it, and the harping voices drawing attention to my many inadequacies and faults would gradually drown out its voice. But until that time, I realized on the trip back north, I had a window, a small and rapidly closing one, in which I could do anything I wanted. And as I turned up the icy mess that was my road in mud season, I knew what that one thing was.

I wanted to walk in the woods.

A month later, I was driving through flat, windswept country on the New York side of Lake Champlain, to the six-hundred-acre Wilderness Survival Center.

I had met Marty Simon, the director of the center, at a weekend retreat sponsored by the Vermont Organization of Wilderness Guides (VOGA), a group that aims to give women a sampler of outdoor skills and experiences. The Doe Camps, as they're somewhat unfortunately called (somehow the image of skittish brown-eyed ungulates doesn't convey competence), provide a crackerjack group of specialists who will teach you how to cook a moose, bring down a wild turkey with a twenty-gauge shotgun, wrangle an ATV, catch a trout with a fly rod, identify edible plants in the woods, and use a chain saw, among dozens of other skills.

Most of the clinics provided tart servings of humility to this participant—the only thing I was able to hook during the fly-fishing lessons was my own coat—but the instructors were nothing if not patient and the women attending were a hoot. It was a variegated crowd from all over the state. There were a few ringers—women whose older brothers had taught them how to do a lot of this stuff when they were kids and were just brushing up their skills and enjoying the camaraderie—but most of us were members of the virgin incompetent, eager

to learn but a bit intimidated by a world that seemed so ada-
mantly male.

There were women who wanted to get back their inner tom-
boys, and women who wanted to understand what sent their
husbands out before dawn to wander around in the cold and
the damp all day during hunting season. And there were a few
women for whom the woods were an exotic place, but not perhaps as exotic as their ordinary lives must have seemed to the
broad majority of those with whom they were sharing moose
barbecue—like LaWanda, a young woman who worked nights
as a nurse's aide in an assisted-living home, raised three children during the day, and had once been on the receiving end
of a bullet in a Newark housing project.

The course I had been most interested in was a two-day
seminar in wilderness survival, taught by a gravel-voiced,
balding, mustachioed Vietnam vet with the stoic eyes of a bas-
set hound, who tossed out jokes, advice, anecdotes, and patter
with the aplomb of a stand-up comedian working the small-
town nightclub circuit.

Marty Simon understood women—he didn't condescend
to them by dumbing down what he had to teach or by treat-
ing them as rarefied beings who couldn't take a joke. He dis-
armed through provocation, making outrageous, non-PC
comments about marriage and women, most of them directed
at his wife, Aggie, who sat at the back with their two dogs,
and with whom Marty locked eyes with such frank admira-
tion that he immediately inspired trust.

In my first few months in Vermont, I had surfed the Internet
in search of a course or a Web site that would teach me what I
wanted to know, and was inevitably sidetracked into reading

countless tales of what had happened to those who went into the woods without these skills. None of it was of much help—there was a not-so-subtle machismo running through most of these tales, a tone of "here we are being manly men, look at the risks we take," that made me want to get back to a long-neglected needlepoint tucked away in the linen closet.

The same mixture of arcane knowledge and pumped-up posturing informed most of the Web sites offering instruction, and the choices were bewildering. The orienteering schools offered to teach you how to use a map and compass in about fifteen minutes before sending you off with a timer to compete in a race to find your way to some hidden cache before anyone else did. It was the woods on Ritalin. Then there were the camps for survivalists that taught you how to field-dress an elk so you would have something to eat after civilization had ended, and on the other end of the spectrum, the quasi-mystic types who had learned their craft from Native American spirit guides only they could see. It was during those online expeditions that I had first run across VOGA's women-only weekends, which sounded a good deal less overwhelming.

Marty started us out in the classroom and then, an hour later, we took to the woods. We learned about fire making—never walk in the woods, Marty said, without a bit of cotton steeped in Vaseline and a flint; together they would spark a fire even in damp tinder. We also learned how to start a fire with a stake whittled from a tree, a bow made from a bit of twine (another backpack necessity), and a supple branch. The process was difficult and well-nigh interminable, but it worked. He showed us the flat, dry high ground free of roots and protected from falling branches that offered the best places to build an emergency shelter for the night, and the places not to, where the dew would seep through your clothes and insects would drive you nuts. He introduced us to unassuming little bits of greenery that turned out to be rich sources of nutrition growing right under our noses, pointing out delicate leaf patterns that grew from white carrot-shaped roots that tasted like cucumbers. Pretty soon we were all wandering around looking for choice spots to spend the night, pulling up miniature albino root vegetables, and choosing the best branches for insulating ourselves, with the enthusiasm of young married couples looking for a new home.

Most if not all of us would have little use for this information in our daily lives, and at first it all seemed a bit Boy Scout. But then I got it. Marty had not only made these archaic exercises fun, but he was also minute by minute instilling confidence that would stand us in good stead in any situation in which we found ourselves. Marty made competence a joy in and of itself, tying it to a way of seeing the world that was dazzling in its practicality. That was when I decided he might be the one who could teach me direction.

At the end of Marty's course, I purchased my very own sparking flint, mostly for the sheer elegance of the smooth, dark gray cigarette-shaped bit of metal. Then I hung around until all the other women had left. As he was packing up, I asked Marty a little nervously if he would consider giving me private navigation lessons. He was wary, especially after I told him I would probably want to write about the experience. He had a very low opinion of the press, and besides, he was too busy, he said, but in the end he gruffly agreed to let me write to him and give him a better idea of what I hoped to learn and why I wanted to learn it from him.

I didn't write the letter for a long time. I was afraid, that was the long and short of it, though I didn't admit that to myself at the time. Afraid of failure, of exposing my incompetence to a stranger, no matter how understanding he seemed. And afraid of something else, something I couldn't admit to. Or couldn't before I had cancer. It turns out that after you have stared out a window and seen a bespectacled egg where your vanity used to be, you are ready to admit to pretty much anything.

And what I admitted to myself when I got back to Vermont and put the yellow roses in water was how very much I had always resisted a sense of direction, how hard I fought to go the wrong way, even in the simplest of situations, even when listening to an automated voice telling me to go left on a one-way street. (Wrong street. Had to be.)

Part of it was fear of any information presented arithmetically or for that matter spatially, as well as the anxiety of getting it wrong, of failing the test a map represented. And part of it was the eternal bliss ninny aspect of my soul that always seemed to prefer magical thinking to reality. But the reluctance went deeper than that, went to the heart of what it meant to take responsibility. I could take responsibility for others—my daughter, my husband, my friends. But to take responsibility for myself was alarming, in a way I hadn't understood until cancer forced a reckoning.

I understood it now. And while I still suspected that my fears about finding my way might be too deeply ingrained and intractably neurotic to reform, I wrote a letter to Marty.

Marty agreed to put me through a two-day course, instructing me in the basics of what is technically known as land navigation. In his reply, he assured me that anybody could learn how to find their way, and it was that offhand assurance that I clung to. But I had my doubts.

By the time I met up with Marty it was late spring. I got lost, of course. Despite Marty's careful instructions, I had turned to my cell phone in a moment of uncertainty, and the phone had led me to a road taken out by a gully washer two years earlier. Marty wasn't surprised when I finally lumbered into view of the rough shed in which he teaches the classroom portion of his courses. It's why he doesn't include GPS in his instruction. People won't follow their own eyes anymore, he said. They're so dependent on technology they can't do anything for themselves.

We tried to get right to work, but the drumming rain on the metal roof over the shed had us yelling to be heard, and we retreated to the house, where Aggie made us tea.

We sat in the kitchen, waiting for the weather to lift.

Marty had mentioned something at Doe Camp about the importance of being prepared at all times for just about anything, so we talked about what he took with him when leaving the house. It was a pretty impressive list: three knives, including a Bark River all-purpose camping knife handmade to his specifications, a Bravo Necker 2 that he wore on a lanyard under his shirt, and a smaller model in a sheath at his ankle, just in case. In case of what, wasn't clear, but I wouldn't want to be the terrorist chihuahua that crossed his path.

The knives were just the beginning: there were also a standard compass, a wrist compass, and a .357 Smith & Wesson stainless steel five-shot revolver, which he was licensed to carry concealed in forty-one states—he was working on the remaining nine. In addition he wouldn't even go to Walmart without a lightweight compressible thermal body bag that folded up to the size of a pocket handkerchief, and of course a sparking flint and a small tube of cotton soaked in petroleum jelly.

All of which makes sense if you believed, as Marty did, that the end of freedom as we knew it was close at hand, that Hillary Clinton, Barack Obama, and the like were in league with the anti-Christ, that there wasn't much difference left between us and North Korea, and that any sane man or woman had a ten-year supply of canned goods and dried meats stored in his or her home or bunker for the coming apocalypse.

It would have been easy to dismiss Marty Simon as a cartoon version of a right-wing gun nut. The reason I didn't was Marty himself. There was a kindness about the man, a sense of integrity and character that prevented such easy facile stereotyping. I would trust Marty Simon with my life, I remember thinking. A grand statement that usually means nothing, and maybe it occurred to me because at that time I was myself in survivalist mode from the whole cancer ordeal. But something about the man made me think about the number of people I knew about whom I could make that statement. The list was extremely short.

The rain began to let up, reminding Marty of why I was there, and we got down to business. People got lost, Marty said, out of a combination of ignorance and arrogance. They don't prepare, they don't look at a map and plot a route, they don't know how far they're capable of walking in a given amount of time and at what pace. They don't bring what they need in case something goes wrong, and they don't pay attention: to the landscape, to the way in front of them and the path behind.

And in the end, they lose control, shutting down the very parts of their brain that they need to survive. You've got to know you, he said. I can't tell you that often enough. And he couldn't, though it would take a long time for me to understand what he meant by it.

Marty divided his classroom instruction into map and compass, first individually, then in combination, before heading out to put them to practical use in the field. With great deliberation he pulled out a deeply creased topographical map of the area in which the Wilderness Survival Center was located, smoothed it out on the kitchen table, giving me one of his sidelong glances that took in the anxiety and fear of failure that must have been radiating from my side of the kitchen table. I could see the thought *this might take a while* crossing his face.

He was right. When I thought about maps, to the extent that I did, it was usually from two very different approaches. The first was simple intimidation—I had never been able to read even the simplest of maps, and my incompetence baffled me. At the same time, I loved maps, the very idea of them, the supreme audacity of what this two-dimensional rendering of the world was aiming to accomplish.

Look at a map and logic blooms, orderly and incontrovertible. There is the world rendered submissive to reason, destinations linear, alternatives manifest. North is up and south is down, east to the right and west to the left. Maps present a comforting vision of the world, reducing infinite distances to a grid composed of precisely drawn horizontal and vertical lines, tethering treacherous winds to a four-pointed star. There is no arguing with a map. Yes, you can get there from here, the map proclaims, and this is the path you must take.

But a map begins as a trick of the eye, a sleight of hand, the round rumpled sphere of the earth made flat and smooth and fitted to a rectangular piece of paper. Already you are in the land of the counterintuitive, and in fact everything about direction speaks to this ineffably human quality of trying to impress a logical sense of order on a chaotic and shifting world. Even the four cardinal directions— north, east, south, west— began as an act of hubris, defined not by the contours of the world itself but by those of the human body: anthropologists believe the first directions were simply up and down, back and front.

There were maps before there was writing, scratched on cave walls, on papyrus and parchment, and from the beginning they have represented more than a simple picture of where things are in relation to one another; the earliest maps pointed the way to paradise. They are works of both poetry and prose, metaphor and explication, of wild imagination and narrowest calculation, a place where faith and reason both collide and coexist, a tracing—in the eyes of Gerardus Mercator, the Renaissance cartographer—of the hand of God.

I don't remember how much of this I tried to impart to Marty in a desperate attempt to conceal my inability to understand what he was talking about, but I knew I was in pretty deep when I found myself telling him about how you should never walk widdershins around a church, because by going counterclockwise, or opposite to the sun, people thought they might enter the land of the fairies.

And not the good kind either, I concluded.

Right, Marty said. Then slowly, as if talking someone down from a ledge, he began to explain the topo map.

At first it seemed a head-splitting and most likely impossible exercise in deciphering hieroglyphics. Topographic maps, as issued by the U.S. Geological Survey, are compiled from photographs taken by the National Aerial Photography Program with specialized cameras embedded in planes flying at precisely consistent altitudes in a north-south direction along predetermined flight lines. It took ten such photographs to provide the stereoscopic coverage necessary and at least five years of calculation and observation to compile the 7.5 minutes of latitude and longitude represented by the map spread out in front of me.

The result is a large rectangle of soft thin paper covered in swirling green and brown lines and dotted with circles and squares in black, blue, and, occasionally, red, which at first bear little or no resemblance to the physical world, but which gradually reveal themselves to represent every natural feature and man-made object that lies between you and where you want to go.

In addition there was a host of numbers to confront. Think of the earth as a giant magnet, Marty said. The movement of the molten outer core around an inner core crystallized by gravitational pressure generates a roughly bar-shaped magnetic field that shoots through the globe from the North Pole to the South Pole—more or less. While true north lies directly beneath Polaris, the North Star, magnetic north, the one toward which the compass points, is somewhat to the left, in our hemisphere. The discrepancy varies the farther west you travel, and map and compass are caught between the two, the map indicating true north, the compass magnetic north.

In my front yard, for instance, the compass's idea of north is 15.5 degrees to the west of true north; in Fairbanks, Alaska, it is about 27 degrees east. Adding to the confusion is the fact that the location of magnetic north moves over time, because of changes in the earth's core: in 2009, for instance, it was still in Canadian territory but heading toward Russia at the rate of thirty-four to thirty-seven miles per year.

Okay, I didn't get all or, for that matter, any of this at first, except in an occasional flash of insight where I could suddenly see what was going on between compass and core, map and star, before the whole thing vanished into murky seas again. But that didn't matter so much as the fact that the entire scale of things had changed: we were no longer talking about some abstract arithmetical problem that invoked my dread of numbers and three-dimensional thinking but of something concrete and physical—immense, yes, but real nonetheless, a planet spinning, a star fixed, a jumpy little needle inclined always toward home. Direction wasn't about what was in my head anymore; it was about the world.

And, gradually, that world unfolded. The mysterious squiggly lines became an intricately detailed panorama of information. Contour lines, the soft sepia undulating circles that are the predominate feature of a topo map, could be translated into specific images of the terrain—where the land was level and easily traversed; where it was steep; the change in elevation over a given distance. The more closely one studied the map, the more there was to see: the seemingly random numbers were actually fixed intervals, denoting elevations—count the number of contour lines you cross and that number will yield the change in elevation. The map could tell you where a river ran too fast to be forded, or was too deep to cross, where a chasm would yield to a less treacherous passage, when to take the direct route and when the long way around was the wiser choice, because those wavy blue lines meant a swamp, and that seemingly inconsequential hill was as steep as Denali. Little black dots were houses, larger ones a barn or maybe a school; logic would tell you which. Red was a main road; purple, a state park; a solid blue, a river; an intermittent one, a seasonal creek that dried up in summer.

It can take a lifetime to master reading a topo map, but the most important lessons are more easily learned, although like most of life's lessons they must be learned over and over again.

One of the skills Marty pushed most relentlessly that first morning at the kitchen table was the importance of orienting the map. This is not a complicated idea: to most people it's probably strikingly obvious. But I found it almost impossible to understand at first, coming as I did from a lifetime of bending the world to my own notions instead of taking it as I found it.

The concept is simple enough: if you were to draw a map of your living room, but you were holding it the wrong way, the map would take you in the wrong direction, no matter how detailed and accurate it was. Farther afield, it's clearly even more important: if you are holding the map with north at the top, because that is the way you can read it, but you are heading south, then you will march off in the opposite direction. To correct for this, you align the two realities so that north at the top of the map is the same direction as north on the compass and you have turned the map until it mirrors the direction in which you're headed.

It was hard going, and we practiced at the kitchen table for a long time; I had a rough time getting rid of the map in my head, the one that said that north was straight ahead no matter which way the map was turned, and trusting instead the one laid out in front of me. In my brain, whatever I'm looking at is north, an attitude I apparently carry over to my friends, who unanimously agree I am one of the bossiest people on the planet (although I've always thought I was something of a pussycat). The rusty hinges of my egocentric certainty took a lot of coaxing to finally loosen up. Practice, Marty advised. It would take a lot of it. But if I could start looking at the world and the direction I was taking as it really was, I might actually end up at the place I had set out for. It seemed too good to be true.

Marty made a point of emphasizing that none of this worked if you didn't study the map before you actually headed for a destination. Acquiring an accurate idea of which way to go was almost impossible without preparation: planning a route, taking note of the obstacles, calculating the time.

To this I objected. From the time I moved to Vermont I had been asking nearly everyone I met in the woods how they found their way in the maze of old logging roads, bridle trails, uncut brush, and meandering stone walls through which they wandered. They always looked at me kind of funny—I guess to them it was a little like asking how they breathed. They just knew where they were, they told me.

That's what I want, I told Marty. I want that kind of sense of direction. Where I always know where I am. I want to be able to read the landscape, to gauge from the sun and the trees where I am, to carry the direction home in my head.

Marty shook his head. I can show you a few things that will help you, he said, if you're lost without a map or compass. But what you're talking about, that's not an instinct or being one with the woods or any of that crap. The people you're talking about, they know the area. They're familiar with the landscape. They have probably been walking those woods for years.

I had to admit that made sense. Marty's words called to mind a line from a documentary I'd seen. Two Alabama hunters are asked how they find their way in the trackless swamps in which they roam. We don't need to know north or south, says one of them. We got the trees named.

And that is the crux of the matter. So much of direction, of having a sense of direction, is bound up in a sense of place, of knowing where home lies even when you don't know exactly where you are. If you have walked a wood for so many years that the trees have names and characters, are by way of being old friends, then the route you take is a kind of songline, a path of memory, mood, desire, regret, of passing moments and striking images that create maps more durable than anything on paper or beeping bits of metal and plastic.

In my neighborhood, it is much the same. When I first arrived in South Woodstock, I was asking directions constantly, and just as constantly coming away confounded. How do I get to the store? I would ask the two guys mudbugging on a dirt road whose name I didn't know. Easy, they would say. Just walk three hundred yards and hang a left at Connor's pond. Head down toward the Fullerton cemetery, but if you hit the Giddyup Road, you've gone too far east.

The Giddyup Road?

Yeah, the bumpy one where those boys crashed their rig that summer. We call it the Giddyup Road.

Ah, yes. Thanks.

But I wasn't from South Woodstock and I would never share that sense of familiarity and long association. The thing I wanted—the ineffable ability to read the landscape, to find clues in the lay of the land, whether you knew the place or not—apparently didn't exist. I felt a little disappointed. Maybe like the ancients, I needed direction to retain a little magic.

The rain finally stopped and Marty and I went back out to wilderness headquarters to practice using the compass in the real world.

I learned how to shoot a bearing, so that I could set a course toward an object that couldn't be seen in the distance. "Pretend your Jeep is a mountain," Marty said, "and tell me what direction you would have to walk to get to it." I stood out in the field, holding the compass level, with its sight line aimed at Mount Jeep. Then I slowly rotated the bezel surrounding its face until the north sign was directly over the north needle, what Marty calls "putting the red in the shed." That done, the small notch in the front of the compass now told me that the Jeep was 260 degrees northwest of where I stood. Eureka! For an instant, the connection between a planet wobbling its way around the sun, a distant star, a thin red needle, and a muddy set of wheels made sense, and I marveled at the elegance of it all. Maybe I could do this.

I spent much of the next day marching up and down a dirt road. Marty was adamant that pacing was instrumental to finding your way: it all started, he said, with the ancient Romans, who devised the first unit of long-distance calculation, the *mille passuum,* or "one thousand paces," each pace consisting of two steps: a thousand Roman paces worked out roughly to five thousand feet, and eventually the Roman *mille passuum* evolved into the English mile.

Pacing, Marty said, was a way of measuring distance in the field: if I knew how long it normally took me to cover a known distance, I could predict how far (or far afield) I had traveled in the woods. I would know, for instance, that if I had traveled for two hours in the direction of a lake and that in that time I normally covered about two miles over rough terrain, then I would know whether or not to be worried if the lake was nowhere in sight.

Marty was right, that was pretty cool, but practicing pacing was boring when it wasn't frustrating. Because everyone's individual pace is different—depending on the length of their legs and their style of walking, two people both six feet tall can have strides that vary in length significantly—it was important to learn your own. You did this by practicing, on different kinds of terrain, and under different conditions—uphill in high grass, downhill on gravel, in the rain and in the heat—learning how your own body reacted and adjusted to changing circumstances, so that eventually you would have a rough idea of your own individual average rate of speed. That Greek guy Socrates knew what he was talking about, Marty said. You're gonna have to learn you.

Marty had laid out a course on the long flat dirt road leading to his camp marked off in hundred-feet increments. With paper and pencil in hand I walked to the first marker, counting my paces in military fashion, leading with my right foot, and counting two steps as one pace. I wrote down the number of paces and then, starting once again at zero, counted off the paces to the next marker and then again to the third. On the return, I counted the paces straight back from the three-hundred-foot marker to the beginning. Then we compared the individual segments with one another, looking to see how consistent the number was.

Eventually, Marty said, we would arrive at the length of my average pace—a number, I gathered from the tone in which Marty explained this, that would rival the square root of *pi* in significance.

And so I paced. First up the road, then down the road. I tried to walk the way I usually walk, and to count—and one, and two, and—without letting my mind wander, but my numbers were wildly inconsistent. It was as if I had minced along in a kimono for one segment and galloped through a strenuous polka the next.

For a while Marty was undaunted, but at last even he gave up. I'm only teaching you the tools, he said. Then you're going to have to practice. He suggested I make myself a set of pacing beads, so that I could tick off one bead for every ten paces I walked, which would give me a rough idea of the distance I'd traveled and, assuming I regularly checked my watch, the length of time it had taken me to get that far. It sounded like a lot of work. Did Aggie have pacing beads? Aggie doesn't need pacing beads, Marty said. She has me.

Finally I was ready for a field exercise in which I would put together what I had learned so far. Marty had a course laid out, consisting of four posts positioned in a rough quadrangle over a large field normally used, he said, for tomahawk throwing competitions. I was to site my compass on each post successively, writing down the compass reading and following it until I could see the next post, while also measuring my paces. Done correctly, the bearings would bring me full circle back to my starting point and would correspond to the ones Marty had measured with a surveyor's instrument. My paces would tell me how long each side of the quadrangle was.

It took me about an hour. I walked with a Zen-like consciousness of every detail of that patch of ground, from the density of the brush, to the quality of the light and the strength of the breeze, as I tried to get the most accurate reading possible on the compass. The hyperattention made the terrain exotic, as if I were walking through heavy jungle or frozen tundra. I felt oddly confident—despite my shaky hold on what I was doing, there was something aside from myself to trust, a set of skills that, if relied upon correctly, would carry me through. In an odd way, it wasn't up to me.

At the end, I compared my compass readings with the accurate ones, and my paces with the actual length of the course. The results were a decidedly mixed bag: my compass readings were pretty good, while my paces, not surprisingly, were spectacularly off. I was excited about the compass, which was beginning to make sense, and while I knew I would never become deeply acquainted with a set of pacing beads, I saw why Marty was so set on my understanding the underlying concept. It was another way of keeping oneself in the present, of staying attuned to what you were doing and how you were doing it in an environment as changeable and unpredictable as the natural world.

We crammed in a lot over my two days at the Wilderness Survival Center and still just broke the surface of the intricacies of land navigation. Working with the map and the compass together, I learned the rudiments of determining my location by aligning the three-dimensional world around me to the two-dimensional one on paper, and how to work the same idea in reverse, using distinctive features and compass readings on the map to indicate which mountain or lake I might happen to be facing.

None of it would be of much help for the kind of navigation I had in mind, Marty said, the wandering about in the woods where there was little in the way of landmarks to act as guides and waypoints. That would only come in time and with a lot of practice, if it came at all. At the end he demonstrated some of the more homespun methods of finding your way—how to assess direction from the angle of the sun on an analog watch; how to determine north and south by the way a shadow falls from a stick planted in the ground; how to locate the North Star and calculate declension from the night sky. But Marty disdained the wilderness types who claimed that such skills were all you need, who believed that if you listened hard enough, you could hear the trees talking. Don't try to be friends with nature, he said. Nature will kill you. I tell people on my Web site, if you're looking for God or a spiritual experience, you're in the wrong place.

I left the Wilderness Survival Center bemused. Direction was not as difficult as I imagined, nor as fundamental as I hoped. Wayfinding was an art both simple and complicated: simple in its individual components, complex in the harmonies that map and compass, time and terrain, broad estimates and detailed calibrations must achieve. Direction was not beholden to grace or instinct; it was a puzzle that yielded to preparation, precision, and patience, to self-knowledge and a large dose of humility in the face of the unknown. None of which I possessed in any measurable quantity, but then, I had time. Sometimes, when you are lost, in life, as in the woods, the best thing you can do is stay put.

Finding my way, Marty had said, came down to practice, not merely because practice would make the now exotic techniques of wayfinding more familiar, but because practice led to experience—not only of map and compass but of the person using them. Learning how not to get lost was about knowing your own limitations, about what you couldn't do and didn't know, as much as it was about the reverse.

I thought about that on the drive home. About the need to factor in one's own limitations when charting any kind of course. The trick was weeding out the insubstantial stuff of fear from the basic elements of who you were, the nature of your own nature, the quirks and distortions in the fabric of your character that made you who you were.

I got lost for many reasons. For one, I had been too lazy to learn the basics of navigation, a neglect I was now doing something about. That should help with the fear. But I also got lost because I daydreamed, and daydreaming was one of the best parts of walking for me. I would have to figure out a way to account for that predilection, to figure out a way to fit my spacey ways into a plan that would also get me home. That's what Marty meant when he said, You have to learn you. Not just your strengths but also your limitations. I was glad he included the latter. It gave me a great deal more to work with.

Heading up the road to Castle Dismal, I also thought about something else Marty had said: to know where you are going, you need to know where you have been. Looking back, fixing in your mind the path behind you, was as important as going forward.

The Wolf Tree

And the seventh sorrow
Is the slow goodbye.

—TED HUGHES, "Winter in the Village"

That fall I made frequent trips to my mother's house. For years, longer than any of her three children were willing to admit, she had been losing ground, following the usual progression of dementia, from the increasing number of misplaced objects to the day she couldn't find her way home to the neighborhood in which she had lived for nearly fifty years. She had taken a taxi to the airport to visit my brother and his wife in Florida. But the crowds at the airport confused her, and she turned on her heel and went back home. Or tried—she couldn't remember how to get there, and the cabdriver spent hours driving her around the neighborhood before she finally spotted a familiar street.

That's when my brothers and I began to talk to her about making some changes. At first we thought that perhaps she was simply too isolated, that she was becoming depressed—she had stopped going to church or participating in the choir or attending the Polish culture classes into which she had thrown herself with such gusto in the years following my father's death; maybe if she spent more time out of the house, she would feel better. Holly, my sister-in-law, took her to visit a community center that provided day care for the elderly. It seemed a pleasant lit-tle place, offering activities for every degree of lucidity, and we thought my mother, an intensely social being, would enjoy the respite from her isolation. Mom didn't say much during the visit, apparently, but afterward, in the supermarket, she gave Holly the slip in the meat section. Holly found her three aisles over, in canned goods, talking to a startled young woman with a small child in tow. Help, she was saying. Help! My children are trying to put me in a lunatic asylum!

Holly and my younger brother Chris lived about forty min-utes away from my mother, and to them had fallen a dispro-portionate share of the task of maintaining the illusion that my mother could take care of herself. It was getting to be too much; I came down that fall to provide some backup. I had the time—the treatment was over, and Zoë was spending a semes-ter abroad in Africa.

Together we tried to suggest to her that perhaps she could use a little part-time help, someone to do the things that increas-ingly were left undone. Like what, she demanded, and in truth, the items on the list, taken individually, sounded rather trivial at first. But over time they had begun to mount up. We mentioned the bathroom toilet that never got fixed, the bills that occasion-ally weren't paid, the dishes that went back into the cupboard unwashed. We mentioned how angry she became when we tried to do them for her. Leave that alone! she would shout. Get out of my house! We talked about the minor traffic accidents, the possibility that she should no longer be driving.

Anyone who has been through the experience knows the black humor, the surreality, and the guilty, pointless anger of these conversations. We were using logic in a land where logic had fled, we were holding on tightly, unconsciously to the parent who had for better or worse been the towering author of our lives. It seemed desperately important to get our mother to recognize what was happening to her, because if she could still do that, then the relationship was still intact. But, of course, for her, it was an admission she could not make without the very ground beneath her feet dissolving.

By November, it was clear that something had to be done. We tried a part-time solution, in the form of a gentle young Ugandan woman, a recent émigré. But Cecilia's endless patience and limitless Christian forgiveness were no match for the torrent of abuse and epithets, and finally physical attacks, my mother hurled at her, nor for my mother's tearful pleas for forgiveness when she was told what she had done. My brother and I began to visit places where she could live full-time.

I came back from one of these trips to find her sitting on the gold brocade sofa in the living room, staring out the window at the silent cul-de-sac. She had lived in the house, a traditional suburban Colonial, for over thirty years, since my father had retired from the military and I had gone to college. She spent most of her time there now in the living room, surrounded by the treasures she had fought so hard to acquire as a young army wife, the painted screens and hand-carved teak furniture, the porcelain vases and ivory statuettes that were the usual booty of such postwar tours of duty in Asia. The meat might be rotting in the refrigerator, and outside, amid the dead leaves on the unswept patio, a line of scorched pots and pans now bore witness to the loss of her cooking skills. But the brief and sudden sunsets of late fall always fell in this room on a scene of spotless perfection.

She smiled when I walked in—a rare occurrence in those days. We couldn't make her understand that we wanted to help her, and she eyed us, fearfully, angrily, as the enemies who would take her away. She was trying so hard to pretend that nothing was wrong. Her paranoia and distrust were part of the disease. They were also well founded, of course: we were trying to take her away from the place she said she would only leave feetfirst.

Don't try to reason with her, said the elder-care experts we consulted. But selfishly, I wanted my mother back, the embattled, abrasive, intrusive mother I knew, not the frightened, failing woman she had become, and so I tried again to tell her about the places we wanted to show her, about how worried we were.

We fell off the usual cliff, of course. Before long she had grabbed a section of the *Washington Post* and a pen. Tell me, she said. Tell me everything I'm doing wrong, so I can fix it. By then I was sick of all that couldn't be said or, once said, was immediately forgotten, so I tried again. Okay. You don't change the lightbulbs! I began, as if it were a felony offense, and as the words left my mouth I began to wonder who in fact the crazy one was.

She looked bewildered. Really? Okay. She wrote in the margins, next to a movie review, I don't change lightbulbs.

Chris took you to the doctor, I said, seeking firmer ground. You have vascular dementia. Your memory is being destroyed. You ask the same questions over and over again, you can't remember what you did an hour ago.

I still have the yellowed newsprint on which she took her notes of that conversation. She wrote in the wavering but still graceful longhand she learned from the nuns at the Polish Catholic grammar school in Pittsburgh. The words, clutching at a present that had already escaped her, crawled along the headlines for a recent photography exhibit on the front page and curled like smoke over the golden anniversaries announced on the back of the paper.

1. Memory Depose Defective.
2. I ask same question over
3. Never.
4. MRI. Long term memory destroyed
5. Cannot drive because broken.

What else? She yelled in the hoarse, strained voice that was all that was left to her. What else. I found myself idly wishing that I could go to Zoë in Africa, to sit her down and read a storybook to her then and there, the one that was her favorite. *Once there was a little bunny, who wanted to run away. If you run away, said his mother, I will run after you, for you are my little bunny.* The memory of those long-ago evenings calmed me, reminded me that I was both mother and daughter, just as the woman in front of me was both mother and child, one who also needed reassurance, as she drifted further and further away to a place I couldn't go.

What else? yelled my mother. What else?

I wanted to pull her frail body close and stroke her balding head and ask for her forgiveness, but there had been none of that in her family, as she put it, and so there had been none of that in ours. Never mind, I said, never mind, let's just watch a movie. I bought ice cream.

So we sat and watched something old and black-and-white and my mother dozed. Finally it was late and I woke her to make sure she went upstairs to bed. By then I had turned off the TV and we were sitting in semidarkness. She noticed I was holding her hand, and she tugged it free the way you do from any kind of trap. You know, she said as she set herself to the task of forcing her arthritic knees up the stairs. Someone should really change those lightbulbs.

Eventually we found a place for my mother. It was a two-story house set on a quiet street in Alexandria, where a maximum of eight people lived out their days provided for by some of the kindest women I have ever met and presided over by an indefatigable former gerontological social worker, Pearlbea LaBier, a thirty-five-year veteran in the field who had come highly recommended by several elder-care specialists. We knew from our own visits that the house was clean and smelled of good home cooking and the residents looked well cared for. There were weekly music sessions and crafts and a shady front porch with white rocking chairs at the ready and the last of the summer roses flowering in the garden. We told ourselves she would be happy there.

We were lying: my mother had never been happy for more than twenty minutes at a time. She ran on fury and worry and the neediness that children who have never been properly loved carry with them for the rest of their days.

She would have been happier if we had taken her into our homes. That needs to be said, because that is the source of the guilt, of course. But none of us could take care of her without that care consuming the rest of our lives. And because we couldn't give her the one future she wanted, I think she was right: she would have been more content if she had been left to her misery in her own home, surrounded by her own things, until she finally fell down the stairs or developed an infection that went untended. But such a course was never an option: instead she would live a longer life in a place she didn't want to be, and that was the best we could do by way of kindness.

I went back to Vermont. I would return to help with the move, just before Thanksgiving, and to prepare the house to be rented; the money would help to pay for her care.

I had time on my hands when I got back to Castle Dismal. The knowledge that the next time I went back to my mother's house would be the last time short-circuited my resolve to get back to work in the intervening days. Instead I waited. I wasn't sure what I was waiting for, though sometimes, in the middle of the afternoon, when the wind was shaking the trees, as if time itself had become tangled in the branches, and energy drifted out the door, I thought I was waiting for the phone to ring. It was the time of day my mother used to call. It had reached the point where she would call a dozen times or more, one call right after another, because she forgot each call as soon as it was made.

Those weren't the calls I was waiting for. I was waiting for her the way she was, when the thread of patience snapped ten minutes after I picked up the phone and the conversation snagged on old rusty pins hidden deep in the fabric of our long, long struggle. When I would end the call speechless with exasperation at how little she understood me, or she would hang up furious that I had failed to understand that all she meant by her criticism was concern, that I never listened to her. But

I had listened too well, until finally, when it came time to take the full measure of my shortcomings, there was no longer a difference between her voice and my own.

The fall colors had been muted that year. The leaves had turned to brown without much ceremony and a hard rain had brought most of them down early, exposing the beauty of the bare black branches and the fact that the apple tree needed pruning. It was an old tree, which produced fruit every other year, good tart red apples, until this season, when the few that appeared were sickly looking, wizened, and small.

At least the apple tree was still standing. I had lost one tree the winter I was sick, in the aftermath of a terrific ice storm. It stood to the right of the house, on the edge of the septic field, an old, bent, broken, gnarled, hollowed grotesquerie of a tree that nonetheless always sprouted a few green shoots in the spring at the top of the blasted escarpment that had been its crown.

The fallen tree was a wolf tree—an old wide-crowned tree with a scarred and gnarled trunk and bent and broken branches twisted into fantastic shapes by decades of fierce storms and lightning strikes. Wolf trees begin benignly, sometimes as sole survivors in a cleared meadow, left behind by a farmer to provide shade and shelter for his animals after the other trees are reduced to stumps. Their solitude protects them for a long while, their size and the width of their leafy crowns, their enormous root systems, stealing sunshine and nutrients from would-be competitors.

No one really knows where the term *wolf tree* comes from, though there are many theories. All I knew was that it suited the grotesque beauty of the tree that stood near my house, and I loved it. I owed it a debt of gratitude as well—when I was new to the woods, it had been the one object I could recognize, the one that told me I was close to home. For me it still stood, much the way the World Trade Center still stood when I searched the southern skyline from the windows of my old apartment in New York. Some landmarks embed themselves in the cartography of the soul of a city, a village, a people. They live as long as memory lives.

I asked a young arborist to come to the aid of my ailing apple tree. Jon Hartland was a diminutive man, with craggy features and a beard that made him look like an elfin Abe Lincoln. He took a look at the apple tree. It needed pruning and nutrients, not to mention better drainage, all of which sounded expensive, and I wasn't feeling too sanguine about its chances.

But the tree was something of an excuse. I was more interested in what Hartland saw when he looked at the trees that surrounded the house and composed the woods in which I walked, what history surrounded me.

Mine was a small patch of northern hardwood, he said, a mixture for the most part of white and yellow birch, sugar maples, and American beech. The forest was young, a mere sixty or sixty-five years old. One hundred years ago, the hills in which my house was tucked were probably home to small sheep farms; when the wool industry failed, the occupants might have turned to dairy farming, or small crops, but gradually the land had been abandoned, and the places where the sheep had grazed were taken over first by goldenrod and asters, which killed the grass, then by brambles and the first wave of pioneer trees, poplars and birches, fast-growing, short-lived trees that gave way to the hardier, slower-growing maples and oaks.

The process was not nearly so decorous as it sounds. In a forest, Hartland said, each tree fights for survival: the trees you see are the survivors of an epic contest, a ruthless struggle for light and air, water and space. It is no wonder, then, if a small current of unease accompanies even the most benign forest walk; around us a battle rages. Hartland pointed to strange twisting branches that veered in sudden verticals or serpentine curves, struggling to reach the light. Some of the trees had even divided against themselves, forming secondary and tertiary trunks, short-term solutions that would betray them in old age when the fissures and fault lines of the past exposed them to wind and weather.

I asked Jon Hartland about wolf trees. After the old tree died, I had gone looking for others. I was drawn to them, to the idea of their ruined fantastic beauty, but another had been hard to find.

Hartland wasn't surprised. Wolf trees are always large, he said, with thick trunks and widely spreading crowns, defined as they grow older as much by what they have lost as by what they retain: the limbs sheared away by wind and ice, the blackened holes gaping like wounds, and the sharpened point of broken branches. The wolf tree's size, its strength, is also its downfall. A wolf tree survives as long as it does because it stands alone: at the top of a hill perhaps, or near the remnants of an old stone wall. But over time, the tree grows old and weakens, and the forest once again encroaches. The huge tree is slowly surrounded by the younger, stronger invaders, whose limbs are supple against the heavy snows and whose root systems are more efficient. The old tree, no longer alone, cannot survive.

A tree dies from the top down. The crown withers, the branches become fragile, the center dries up and hollows out. Still, the tree will stand, to all outward appearances alive, like an ancient warrior brandishing his weapons against all comers. But not forever: eventually it surrenders, to a hard summer rain or a biting nor'easter that knocks it down, and the enormous root system nearly as wide as the tree was tall shuts down.

I wouldn't find wolf trees on the ridge where I had been looking, Hartland said, because the forest itself was still young, although its character had been formed eons ago.

When Hartland drives down U.S. Interstate 91, where it parallels the Connecticut River, New Hampshire on one side, Vermont on the other, he can trace a catastrophic tale of two broken drifting worlds. About 420 million years earlier, he said, New Hampshire and Maine had been a part of the African tectonic plate while Vermont formed a part of its European counterpart. Just east of Vermont 100, the old road that runs roughly down the middle of the state, the African plate crashed into the European, obliterating what was then the coastline, and creating mountains that were eventually worn down by ice and storm and time to the present-day ridges and hills.

Because they had come from different continents, the bedrock and the soil in Vermont were very different from those in New Hampshire a few miles away. New Hampshire had the sandy soil of its African origin and became home to oak, white pine, and hemlock. Vermont, formed at the bottom of an ancient ocean, contained limestone, which was itself the detritus of sea creatures; the higher pH levels of the soil made it home to different species, to beech and birch, sugar maple and ash. Because the soil was fertile in Vermont, things grew and grazed there; because the soil was not in New Hampshire, that state turned to the mills and the canneries.

The mountains and the glaciers that pushed down from the north shaped history, determining the direction of commerce and culture that settled in the valleys and followed the rivers down to the sea. The glaciers sculpted the lakes and carved the mountains, pushing boulders and the remains of exploded volcanoes ahead of them, shaping the topography in an asymmetrical direction, the south side steeper, darker, denser than the north, determining the steeps and flats of the trails I now walked.

You can't think about the forest without thinking about soil, Hartland said. You can't think about soil without thinking about bedrock. And you can't think about bedrock without thinking about time.

The last morning my mother spent in her own home, I made her French toast. She didn't know it was her last morning—we had told her what Pearlbea had told us to think of as "a loving lie," that she had to leave the house for a little while until we got the furnace fixed. It was true that the furnace was always going out and my mother was deathly afraid of the cold, so the lie had worked, though I doubt it would have been so successful if I hadn't also slipped a Xanax among her daily vitamins.

"We eat the same way," she said suddenly. It was startling to hear her make any fresh observation—for over a year her conversation hadn't deviated from an endless loop of a few predictable sentences. In the morning it was an inquiry about how I had slept. In the car, during the day, it was about the vapor trail of a jet going through the sky and how pretty it was, and in the car at night it was about the colored taillights of the other cars and how pretty they were. It reminded me of what being stoned was like, and I sometimes wondered if any of this terrible process was ever pleasant for her.

She was right: we did eat the same way, both of us cutting up our French toast into little pieces and then eating them from the center of the plate outward to the edge. I had spent half my life trying to be anyone but my mother, but by now I was reconciled, at least somewhat, to how much I was like her—the best of me and the worst had come from her. And I thought of how, as she receded into shadow, she was taking a part of me with her, a part of all three of her children. There was no one who would ever see us as capable of such perfection as she did, no matter how religiously we disappointed her.

The residents were eating lunch when we arrived, and they made room for her at the table. She was always happy in company, at least at first, and we slipped away as they made her welcome, unpacking her clothes in the bedroom that was now hers, scattering photographs of her husband and children and putting in its new place the rosary that was always next to her bed.

Afterward, my brother and sister-in-law and I went back to my mother's house. It was so strange: everything was exactly the same as it had been a few hours before and everything was entirely different. It had been a destination, to run away from, to creep back to, and now, still standing, it was gone.

We spent the next few days throwing out an unspeakable amount of trash. My mother saved everything—empty take-out containers and the old wooden cylinders on which sewing thread used to be wound, bolts of beautiful woolen and silk fabric, now stained and moth-eaten, mimeographed home-work assignments from her years as a teacher, every kitchen utensil she had ever owned.

And there were photographs. Boxes and boxes of photo-graphs that we dragged into the room my mother had called the study. No one ever studied there, or so much as entered the room, except for my father, who read the paper there and found a refuge from his wife. It was a pretty little room with in-tensely uncomfortable tufted furniture and dusty bookshelves holding hardbacks from their years as members of the Book-of-the-Month Club half a century earlier.

Holly and I sat on the floor, going through every photo, tossing duplicates and travel pictures, saving images of my mother's childhood, those of her children and grandchildren. There were thousands of them, every family visit catalogued, every ceremonious occasion. We sat there for hours, watching the past unspool, including our own as young mothers. It had been easy of late to tell myself that Zoë's childhood had gone by in a flash, but here was the warp and weft of it laid out stitch by stitch, the ticking minutes, the endless afternoons of feeding and changing and reading and entertaining, of soothing and humoring and scolding and doing it again and again. Wow, said Holly. Life is long.

We packed up most of the pictures to be stored in her base-ment—we told ourselves our children would want to look at them someday. But I took one with me. It was a picture of my mother, an eight-by-ten black-and-white in one of those paper frames that can be propped upright by means of a bit of angled cardboard.

I had never seen it before. A school portrait, probably, taken when she was about fifteen or sixteen. She is dressed in a neatly pressed white blouse with a rounded collar and a woolen vest with bright brass buttons, a jauntily folded handkerchief positioned just so in the left breast pocket. A crucifix gleams, caught in the camera's flash. Her hair falls in tight curls just above her shoulders, having been no doubt twisted up in rags all night for the occasion. My mother had told us for so long that she was ugly that I had believed it until I was old enough to judge for myself, but here was the beginning of her unruly irregular beauty—pale skin, delicate arcing eyebrows, the large nose and the full underlip of her mouth for once not compressed into a hard thin line. And here, too, was what might have been, had she been loved, had she been told she was pretty, had anyone believed in her. There was hope and excitement in her eyes, such a willingness to please and to find pleasure, a belief that there was a place for her in the world as she then imagined it, if she only looked hard enough. There had been no love in her harsh young life: she was the daughter of Polish immigrants, a steelworker and a charwoman. Her mother had been the eldest of ten children who had grown up in a brutal and violent poverty and who considered it enough to put a roof over her children's head and food in their mouths. My mother worked hard, put herself through night school, and hoped for happiness, but by the time it came along, in the form of a young New Hampshire soldier, the damage was too deep: it had divided her from herself. But there in the picture was the woman she could have been, on the verge of being eclipsed by the woman she would become. In the photo I could see both of them, looking out from frightened hopeful eyes, her tentative uncertain smile.

My mother worried that there was something deeply wrong with her. That she didn't belong—because of her poverty, her looks, because her mother declared as much. She was and would always remain an outsider. It's a look you see in the firstborn of many immigrant families—sometimes it's erased by success and happy endings, but sometimes it hardens, and moves glacierlike down the slope of a life, and into the lives of the children who come later, transforming the soil in which they will grow up, determining which side of the ridge they will walk.

I took the picture home and put it on the mantelpiece above the woodstove and took it down again and stared at it for hours, this picture of my mother with all her dreams intact, yes, and with all her fears already formed. And when it comes time these days, as it does all too frequently, to think about what I could have done differently, I try to remember to look at that picture, because it reminds me to remember how much of what happens to us now happens because of the vagaries of an- cient glaciers, because of the way in which continents collided in generations no longer remembered. You did good, I tell the picture, phrasing it the way my grandmother would have, had she ever said the words my mother longed to hear. You did the best you possibly could.

At first, after the confusion of her initial few weeks in the new place, my mother seemed happier to us than she had been for a long while. She had always liked people, but her difficult nature—opinionated, oversensitive, intrusive—had driven them away. But at her new home, she was surrounded by women determined to make her feel loved, and by others too out of it to notice her contrariness. And for a time she grew animated and participated in the word games and the music classes. She took under her wing a patient whose own dementia had robbed her of speech, holding her hand and encouraging her. She told the women who cared for her that she loved them and tried their patience with endless stories about her children—when she wasn't excoriating them as they tried to bathe her.

But when her children came to visit her, the light would go out of her eyes, replaced by a look of watchful mistrust. She refused to walk, although she still could, and she seemed to find it difficult to put a sentence together—at least for us. She was better, the staff members said, when we weren't around, though they were too kind to put it that way.

During one visit she did make the effort to tell us something. We have to talk, she began, but she couldn't remember the end of the sentence, or what it was we had to talk about. My brother tried to steer her away from attempts to articulate the agenda she couldn't remember.

So how are you, Mom? he said. How's life?

My life, she said, is blackness and mold.

Oh really? said Chris, thinking that she was remembering some long-ago problem with dampness in the basement. And what are you going to do about that?

She looked at us, stunned. You asking me that is like a slap in the face, she said. And that was when I knew that she wanted back her solitude in all its burned and blackened beauty, and that no matter how dark the sea into which her past was slipping, no matter how little she knew of the present, or thought of the future, my mother would always remember, until she remembered nothing else, that we were the people who had betrayed her.

About a year later, I found another wolf tree. A cellar hole, so overgrown I nearly fell into it, and a tumble of old stones that looked as if design, not accident, had placed them there, made me stop and look around. The tree stood in what must have been a corner of the pasture. It was an ungraceful old survivor, lopsided, the stumps of its bony branches poking at the sky like fingers, like an old preacher in a threadbare frock coat, half mad with the prophecies no one stops to listen to. I took pictures. I walked on.

My mother lives in twilight, no longer the hectoring, insistent, urgent woman she was, storming our lives with the love and the fury that nearly drowned us all; neither bane nor blessing, not arrow, not anchor. I'm not sure she makes any more of my presence than the tree does, though sometimes I think I spot an errant gleam of recognition, which quickly fades. And when I leave the pleasant little house where she now lives, there is always a moment of utter disorientation—I don't know where I have been, I don't know where to go—and for that moment my mother and I are finally in the same place.

The Harriet Line

It is not down on any map; true places never are.

—HERMAN MELVILLE, *Moby-Dick*

Sometimes, you get lucky. A key falls into your lap, opening a door you didn't know was there.

The DeLorme Map Store's three-story revolving globe can be seen from Interstate 295 day or night, a nearly irresistible landmark to anyone traveling near the exit for Yarmouth, Maine, and a strangely reassuring presence to the congenitally lost—no matter where you are, somewhere on that expansive presence is home. A *Guinness Book of World Records* holder, the globe has a circumference of 129 feet, a surface area of 5,313 square feet, measures over 40 feet across, and weighs over two and a half tons.

Inside a glass atrium at the front of the store, the globe is girdled by three balconies at various levels, and so dominates the place that all the stuff for sale—state atlases, gazetteers, street maps of cities around the world, interactive mapping software, compasses, pencils, stuffed toys—seems almost an afterthought.

I had driven by the store frequently on my trips with Zoë, but I had never taken the time to go in. On the day I dropped her off for the fall term of her senior year, however, I realized that the opportunities to do so were drawing to a close.

Besides, I had a purpose of sorts. I needed, or thought I needed, the latest in topo map software, hoping to get more up-to-date information about my bit of woodland than my old paper copy had to offer. I had stared at its runic mysteries for hours before I understood how out-of-date it was: there were houses now and trails that hadn't existed when the map was drawn. I had learned a lot about direction since I first consulted that now creased and battered bit of paper, but apparently I still clung to the notion of a map's divine infallibility. Maps never lie, my friend Duncan, a brilliant sailor and outdoorsman, always thundered when his wife, Megan, or I would make some stupid navigational mistake. But they did get old.

I explained what I wanted to the pleasant-looking woman behind the counter. Got a compass? she asked. Yes, but not with me, I said. Know how to use it? Sort of, I said.

She walked out from behind the counter and pulled one of the store's simplest and least expensive compasses off the shelf and wrested it from its plastic nest. Okay, she said, handing it to me. Shoot me a bearing to that sign in the parking lot. I did so, more or less correctly, by my lights, in that I was only 180 degrees off, having aligned the magnetic needle with south instead of north.

Now show me what direction a bearing of 225 degrees is. That one I got right—it was southwest, and I beamed when she awarded me a nod of satisfaction. Apparently DeLorme was very picky as to whom it would sell its products. So about that software, I said. Maybe you could just point me in the right direction?

Judy Gilbert, for that was her name according to the lam-
inated tag on her shirt, gestured vaguely at a couple of rows
of shelves containing handsomely packaged boxes. "But that's
not what you need," she said. "What you need is practice."

Gilbert was a registered Maine wilderness guide and had
been an outdoor enthusiast all her life; she still retained the
ruddy enthusiasm of a woman who would put on her hiking
boots and head up a mountain at a moment's notice. She took
seriously the responsibility of saving neophytes like me from
their own ignorance.

I tried to explain the problem: how I lived surrounded
by thickly wooded hills, how I had this dream of wandering
through them at will, how every time I tried I ended up lost
and confused, unable to understand what the compass was trying to tell me. I know, I should stick to the trails, I concluded,
expecting a dry concurrence.

Not necessarily, she said. And then Judy Gilbert changed
everything.

It was a simple thing, really. Start small, Judy said, as small
as you need to feel safe. Mark out a space for yourself with clear
landmarks, so that no matter how far you wander you will run
into one of them eventually. The landmarks you choose are
panic azimuths, or safety bearings, the idea being that if you
walk in a straight line in any direction, you will get to one of
them.

Then, she said, within the boundaries you have set for yourself, figure out a place you want to get to and, using your compass, determine what direction you need to travel to get there.
Follow the route you have chosen until it does or does not get

you where you want to go. Do it again and again, gradually enlarging the area, but always staying within a set of known boundaries, one in each direction, that act as a sort of safety net for your mistakes.

Gilbert then sold me the expensive software I wanted, which, thanks to her, I would never open.

I started small, really small. The boundaries I set were ones even I couldn't ignore: the walls of the first floor of Castle Dismal. I stood at the windows, leveled my compass, and took rough bearings for the four cardinal directions. Then I began to draw a map of the room on the unruled side of a piece of eight-by-ten-inch printing paper.

The northern wall contained three windows that looked down upon the steep hill and the little creek below; I drew the bare winter branches, including the white birch that saved me during chemo, and the twisty parabolas of the creek and the top of the hill barely visible in the distance. The eastern windows, the ones that spilled the morning light onto the sofa where I worked, faced a hummocky clearing under which the septic tank was buried, a pillowy meadow of snow in winter, a wilderness of weeds and goldenrod in summer. I drew the window, framing in a red fox who used to pause there sometimes near sunset. Next to it I drew the eighteenth-century pinewood spindle that hung on the wall, its wooden bobbins bound by fraying yarns of grays and white and browns.

The front door, which in true Vermont fashion was never used, faced south. I drew the doorframe and the view it looked out on: the small yard, the fragile apple tree and tattered lilacs, a thin line approximating my road, the treacherous thread of dirt and rock, of mud and ice and car-swallowing ditches that wound down to Noah Wood Road and connected Castle Dismal to the outside world. On the blind side, the windowless wall facing west, I drew the fireplace and the woodstove that stood inside of it, and the ever-hungry, half-empty woodpile to the right. There it was, the place I had lived in for over three years.

When I was done I held the map the way you always hold a map, with north at the top, and looked out the window from the vantage point of the sofa. The sheet of paper nearly shrieked in my hand. Then, like a clumsy toddler, I turned the crude little drawing around until the birch tree I'd penciled in was aligned with the real one outside the glass. And suddenly, finally, I got it. By way of discovery it wasn't much, it wasn't Newton under the apple tree, or Archimedes in his bath, it was just a middle-aged woman standing in her living room, turning around a white rectangle and understanding that the world did not revolve around her. But it changed everything.

A map demands you do one of two things. Either you ignore it and the world it represents in favor of the picture you have in your head, or you learn to put yourself aside, and fit the image to what is actually in front of you. And if you are paying attention to the difference in scale between the world outside your window and the one you have just drawn, you might even smile just a bit at the dust mote that has settled on the paper, because it marks the relative size of the space in the cosmos you actually occupy.

Great doors turn on small hinges. I fiddled with a few rough doodles until a clump of cartoon trees faced the forest they attempted to represent, and found my place on the map.

Then I stood in the center of the room, near the staircase that divided the space in two, trying to imagine the newel post of the banister as the foothills of a forbidding mountain range that cleaved an unknown and undiscovered country, forcing me to rely on map and compass to get to the other side. Keeping in mind that I would have to make a set of ninety-degree U-turns to get around the staircase in order to stay on the same bearing, I headed for what lay beyond.

My bearing took me straight to the mantelpiece, just as my map reckoned it would, and there were the three ceramic candlesticks, a delicate blue scrolling hand painted on a bone white background, that my oldest friend had given me nearly fifteen years before, and I thought of our buoyant bumpy friendship. Next to them stood a blue pumpkin-shaped enameled box I'd taken from my mother's house when we dismantled her life, and next to it the photograph of my mother that had become such a touchstone, and this time I saw in its hesitant, fearful smile traces of my daughter, and of me. Another calculation took me by the same process to a framed map hanging on the wall, of the small town in New Hampshire where my father grew up, and I ran my finger along the outline of the lake where we once stayed for two shining weeks and the small cemetery where his father and brothers lay, all veterans of one war or another. A few degrees north took me to the ugly green La-Z-Boy recliner, and I thought about those first few days in the new and empty home and the way my footsteps echoed.

I had the hang of it now, but I didn't stop. I explored the room as if I had never seen it before, as if it had taken years to get there (which, in a way, it had), by camel, riverboat, and brigantine (which it most definitely did not). I was in an old world made suddenly new, mapping a small room holding a long life, the dusty bric-a-brac transformed into waypoints, crossroads, and signposts, pointing a way forward, revealing the way back. That is what direction is, after all, a way of seeing, really seeing, what has been there all along.

Now it was time to practice in the real world. I needed a destination, both familiar and unfamiliar, one that I knew how to reach by road, but not through the woods. It had to have boundaries, so that no matter how off course I was, I would eventually arrive at a recognizable border—assuming, that is, that I walked a straight line as dictated by the compass, rather than walking around in circles, which would be my first instinct. There was really only one choice, the one I thought about every time Henry and I trudged down one road and then up another to visit Harriet and Dean, the one Chip Kendall had first told me about—the route through the woods from my house to theirs.

I looked at my topo map. Harriet's house lay almost due south of mine, a short distance away, and that distance was reassuringly bounded by my road, Noah Wood, the creek behind her, and another trail on the other side of the ridge. I oriented the map on the dining room table and looked for the small black square that would be the exact position of Harriet's house. It didn't exist—like my own, the house hadn't been built when the map was drawn. No matter, I would estimate. I made a black dot, drew a connecting line from my house to hers, and laid down the compass: the Harriet Line, I called it. I noted the bearing—160 degrees, SSE—on a piece of scrap paper.

A little while later I had gathered the compass, the hiking sticks, and the dog and walked a short ways down the road, looking for an opening in the stone wall across the way. Then I scrambled across a muddy ditch and into the woods, took out my compass, and looked for the piece of paper on which I had written the bearing. It was back on the dining room table. No matter, I thought, I remembered it.

By this point I had committed about five major sins against the sacrament of land navigation. I didn't really know where I had begun—the starting point was a guess based on a map that was itself outdated—the direction in which I traveled was a memory of one of several routes I had calculated while peering at the map on the dining room table, and my attempts to keep to a consistent bearing were laughable: it was only after about thirty minutes that I realized that the proximity of my metal hiking poles to the compass was wreaking havoc on the bearings I was getting. I hadn't estimated the distance between the two houses and had never figured out what my average pace was. All of which meant that my attempt to get to Harriet's house might have been construed by a navigational expert, or for that matter, the average Cub Scout, as a dismal failure. But to me, that first foray was a thumping success.

For the first time that I could remember, I did not follow my own instinct as to where Harriet's house ought to be according to the dictates of my own permanently skewed mental map. No, this time the mistakes I made were logical ones, based on facts—incorrectly interpreted, but facts nonetheless. I knew my bearing was off because I remembered what the map looked like, even if I had forgotten to bring it, remembered that my route should not be as steep as it was, that I was headed up the ridge, not around it as I should have been. I had an image of the gently undulating lines that indicate a gain in altitude; getting to Harriet's house had not involved intersecting these lines the way you would in a straight march up rising ground, but crossing them at a much gentler angle than the one on which the compass was currently insisting. For a moment I was able to see in my mind's eye the L-shaped angle of the two roads I normally took to Harriet's house and the unknown terrain that lay between them. And that brief vision pointed toward a different way.

So I ignored the compass, although I kept it out as a kind of moral checkpoint, and walked along the slope of the hill at an angle that felt right. Before too long, I had crossed a stone wall—no idea whose stone wall or why it should be where it was -and entered the checkerboard maze of maple trees strung with waist-high plastic tubing.

Eventually I came to another stone wall, and on the other side was something familiar: the wide green meadow that bordered the woods and descended to a largish pond, which in the summer lay placidly in a riot of dark blue lupine. I knew that pond, and was always grateful when I saw it. I thought about swimming there when I was bald and wondered if I would be embarrassed if I were to face the same situation again, wondered if I would have to find out: the kinds of questions that put all the others into perspective.

The pond was across Noah Wood Road from Harriet's house. I had done it: I had not arrived the way I intended, but I had arrived. I am not sure I have ever been as satisfied with something I had done, at least not in this quiet, proud, and private way.

The next time I walked the Harriet Line, I was almost non-chalant. The taste of success from the first outing still lingered, and I was so sure of myself that I brought the compass more as an accessory—this is what all the cool outdoor types are wearing—than for guidance.

Two hours later I stumbled down a steep hill and into an empty chicken coop in the backyard of a home I didn't know existed in a part of the woods I'd never seen before, convinced I had walked over at least a half-dozen ridges and was probably in New Hampshire. In fact, I was about a quarter of a mile up Noah Wood Road from the Goodwins', a fact I wouldn't discover for another hour, since, not recognizing the road for Noah Wood, I walked back into the woods in the opposite direction.

Most of the people who are reported lost in the woods are hunters, according to search-and-rescue groups. Mostly that's because hunters spend more time in the woods than any other group, but there's another reason as well, and that's arrogance, say some members of the search-and-rescue teams that go out looking for them. They think they know the woods better than anyone and pay more attention to the deer they're following than to where they're going. Arrogance, according to Marty, was the reason most people get lost.

I'm not sure one successful walk had ratcheted me all the way up to arrogant in the spectrum of folly, but it sure did make me full of myself, as my grandmother would say, convinced I knew what I was doing despite very little evidence to support that notion. Besides, my sense of where Harriet's house ought to be was still fighting the evidence of both map and compass and, the evidence seemed to say, winning.

I had started out at the old maple tree at the side of the road, just to the left of Castle Dismal, a landmark I would remember. This time, I wasn't content to merely get to Harriet's house; I wanted to get there by following one bearing, one single point on the compass, which, if followed without deviating, would put me at the Goodwins' front door.

My plan was to use point-to-point navigation. The idea is simple: you head toward the farthest object you can see that lies along your route of travel, and when you get there, you pick out the next farthest thing, until your destination is in sight. It's like a game of leapfrog.

Point-to-point navigation works best in open spaces, with clearly defined markers. In the Australian outback, for example, early explorers used the mounds of the compass termites that grow asymmetrically along a north-south axis as their point-to-point markers, just as the Inuits steer by the consistent direction of the sastrugi, the snow ridges, and the Bedouin by the sand dunes. But in the woods, it's more difficult—the line of sight is much shorter. Since there is no obvious mountain or cleft in the horizon to aim for, you pick out an object that distinguishes itself along your direction of travel, which in a thickly forested area can be only a few yards away a distinctive tree, or boulder. Once you get to it, you pick another on the same directional bearing.

What I found out on that excursion was how important it was to distrust your instincts as a matter of self-protection, because it is so easy to deceive yourself: a compass, for instance, is only as reliable as the hand that holds it. And if that hand is attached to a brain still welded to its own idea of which way to go, it can turn into more of a Ouija board than a reliable indicator. My hand, I noticed, did not stay true when I took a bearing; it veered just a little to the right, taking me with it, leading me downhill. Or a vagrant thought would draw my attention away from a necessary landmark—there were an infinite number of ways, I learned, to lead myself astray.

A wary attitude would have been old hat to a Jane Austen heroine—they knew a thing or two about the tricks played by mind and heart in the Age of Reason. But it was new to me: the vaguely New Age self-help books I'd been consuming like brownies straight from the oven told me to always trust my instinct. What I had been learning over the last few odd, solitary years, however, was that it took a long time to distinguish the small, clear, quiet voice you heard in the soul's silence from the fickle winds of immediate desire—for comfort, for love, for the blunting of anxiety or fear. In that way, my instincts lied; they always had.

I was learning to mistrust those instincts as I was beginning to mistrust the voices from the past whispering just under the radar, retelling the old stories, recalling the old sins, determining my actions or justifying them. They were all familiar companions now, the harsh judgmental voices, the detritus of childhood so well embedded in most of us that we mistake their censure—and their flattery—for the truth. But I was seeing now how often they had sidetracked me, like the unconscious bend of the hand that held the compass.

Those walks along the Harriet Line, learning how to bushwhack, were like a bucket of clear cold water thrown on all that muzzy thinking, forcing me to check myself, to pay attention, to trust logic and calculation and what I could see in front of me.

Still, I made mistakes, a lot of mistakes, on my daily attempts on the Harriet Line, ending up near or behind or in front of Harriet's house (or more often, the chicken coop, which I never aimed for but which always seemed to appear of its own accord). Finally I brought in reinforcements in the form of Hunter Melville, a former Life Scout and current troop leader of Boy Scout Troop No. 20, among his many other accomplishments.

I had met Hunter, his wife, Jessica, and their dog, Laddie, in the aftermath of the havoc caused by Hurricane Irene a few months before. Hunter and Jess had lived in Vermont most of their adult lives. Jess had been a nurse and Hunter had worked at a local ski resort until he morphed into one of those fabled creatures I'd read about but didn't really believe existed: a dot-com millionaire. He and an old high school buddy had come up with an idea for an online vacation home rental service, and he was worth about a zillion dollars according to local report.

Hunter now divided his near boundless energy among a dozen different causes that ranged from Libertarianism to the local historical society In his spare time, he said, he liked to bushwhack. I told him about my own attempts, and he offered to walk the Harriet Line with me and to see where I was going wrong.

It was a spanking bright Saturday in November, and a crisp wind fluttered the loosening leaves. Hunter and Laddie showed up wearing bright orange bandannas around their necks in deference to the hunters—it was deer season. He also pulled out an accessory I hadn't expected—a handheld GPS. But that's cheating, I protested. Hunter just laughed. For him, the gadget was a toy, not a crutch. It was like a nuclear physicist using a calculator—it's not as if he didn't know the stuff already.

We started out on the same bearing I had been using—at my request, Hunter temporarily stowed his gadget and relied on the map and compass to make his own calculations. I was thrilled that he came up with the same bearing, even though we ended up at the very same chicken coop that was beginning to haunt my dreams. The problem, it turned out once Hunter took some readings on the GPS, was that my estimates of where my house and Harriet's were located had been just enough off to put us higher up the hill than we should have been.

We screwed up, I said. No we didn't, Hunter said. You arrived more or less where you wanted to be, he said. What's wrong with that?

I thought about it. There was nothing wrong with that. Hunter's easygoing attitude underscored how I was turning my efforts to get oriented into the same fire-and-brimstone religion I applied to any goal in my life, defining success too narrowly, finding failure everywhere, and assuming eternal damnation was the consequence of landing on the wrong side of the razor's edge that separated one from the other.

The return trip led to more discoveries. For Hunter, any hike was a lesson—in history, geology, economics, depending on what he was looking at. From him, I began to learn how to read the landscape. A cleft between two hills, for instance, indicated a watershed; what lay between them was probably a stream or a river, and if the hills were big enough and the waterway between them was wide enough, then a road had probably been built there as well. Roadways followed waterways as surely as deer trails had led to footpaths worn by Native American hunters, which in turn were widened by the colonial settlers who followed them. If you were lost, Hunter said, the shape of the land itself could lead you to safety. On the way back through the woods, he didn't even need the compass to figure out where we were—he had been noting the angle of the ridge we were traversing and its relationship to the one behind my house.

He pointed to a subtle notch in the hills far ahead, and the dip in the tree line ahead of us that mirrored its shape. Your house is probably there, he said. He had remembered that there was a stream behind my house and hills rising steeply on either side—the dip we were looking at was where the stream cut through. We followed the slight depression in the tree line, heading toward its lowest point, and before long we stepped around a stand of young pines to see directly in front of us, on the other side of the road, my house.

We mystify what we don't understand. After Hunter left, I stood outside and looked around at the woods in which I had walked so often. For the first time, I saw them not as a frightening maze from which only constant vigilance could deliver me, but as a text that needed to be read on its own terms.

I built up the fire and looked at the topo map. Now I could identify some of the features Hunter had pointed out, and the information the map had contained all along sprang to life. Outside, as the day darkened, the woods retreated back into their old atavistic spookiness. Which was reassuring; Hunter had given me a key to the mystery without the mystery itself vanishing. The clues were there if you knew where to look, but

their message was easily misunderstood if you weren't careful to observe accurately, to see what was really there and not
what, however unconsciously, you wanted or feared to see. Like
most things worth knowing, finding your way in the woods
depended on equal parts curiosity, skepticism, and the knowledge that comes only from a willingness to make mistakes. You
had to be able, in other words, to see the forest from the trees.

I thought about Hunter's insouciance as well, his confidence
that the world around him made sense, his ability to enjoy
himself while getting something mostly right, if not perfectly
so. A lot of that had to do with his superior knowledge of map
and compass, and the hours racked up putting them to use, but
there was another lesson to be learned here as well: lighten up.

Except for my bushwhacking attempts to reach Harriet's
house, I still walked the old familiar bridle trails, the ones that
took me through the woods safely and met up with roads whose
names I knew. But one morning, as I was walking past Therese
Fullerton's pond, a stray shaft of sunlight illuminated a glade
deep in the woods. I had noticed this spot before but had never
explored it, too worried about getting lost. Now, armed with
my compass and an attempt at Hunter's nonchalance, I took a
bearing and went off to investigate.

It was a scruffy patch of woods, thicketed by dead pines with
twisted trunks and strangely tinted birches, splashed green
and gray and gold with lichen and age and shadows. I kept
going until I found that patch of sunlight I'd glimpsed from
the path, a small clearing carpeted in a brilliant green moss.
It would be cool in summer; the boughs of the trees inclined
inward and in full leaf would make a thick canopy for the spot.
In winter, it would provide a bit of protection from the wind. I

lay down on the spongy ground cover and looked up at clouds tangled in branches and the tigerish gleam of the yellow birch bark, while Henry poked at a hollowed tree and then ran like hell when something poked back. What is it I lack? the writer and monk Thomas Merton would ask himself, when doubt or desire tugged at his peace of mind. I lack nothing, he wrote, was the answer that always came back. I thought I understood; as long as there were moments like this in the world, it was enough.

When I stood up again, and looked around, I couldn't find the spot where I had entered—the little clearing was roughly circular, and the trees, which had seemed so distinctive as I passed them, now looked like all the others. Normally this would have been the dear-God-what-have-I-done moment, but this time I knew, because I had checked my compass, that the trail lay due east.

I called Henry and headed more or less in that direction, zigzagging a bit whenever something caught my attention. I was beginning to understand how you could both wander in the way I loved, enjoying the brash orange of a tiny wildflower growing in a patch of icy mud, and yet keep your bearings. Knowing that you could find your way back took away from the adrenaline rush of complete disorientation, but it diminished the terror as well, and that was a trade-off I could live with.

I found my way back to the road, though I had made a few mistakes—I hadn't paid proper attention to landmarks and I hadn't taken note of the terrain, whether it was rising or falling, steep or level. And I didn't end up at the exact spot where I left the path. But that was okay. We never go back to where we began, and we probably wouldn't want to.

I kept walking and came to a crooked little side trail I'd never noticed before. It was late and I was cold and tired, but still I wandered down the path, curious to see what lay around the bend. Why does the path ahead beckon so irresistibly that you follow it even when you know you should turn back? Is it only the crooked paths, when you can't see what happens next and you need to find out, as you do in a story? Or is it any road you haven't taken or don't remember?

When I was about to begin chemotherapy, a dear friend told me she wasn't worried about how I would manage. "You'll do fine," she said. "You're good at catastrophes. It's just normal life you screw up."

I think that's true for a lot of us. It's easy to navigate when your destination is clearly visible, no matter what obstacles lie in your way. A simple matter of optic flow, the experts tell us, and the same neural circuitry that gets a beetle to a blade of grass will suffice. You do what you have to do, walk through the brush or around the swamp, get through the chemo, bury the dead, deliver the child to the far side of senior year. When you have the object in sight, you just mow over the bushes. Because when things go wrong, the adrenaline kicks in, and then you are thinking with disaster brain instead of normal brain.

Disaster brain is invincible. Disaster brain says, To hell with the neurotic self-doubt! Who cares if you're no good at anything? You still have to get through this mess you're in, so just do what you have to do because nothing else is important. Disaster brain is cool.

Normal brain, on the other hand, is not cool. Normal brain stumbles around in a swamp of negative thoughts: What if my life has been a giant mistake? What if I really am doomed to be a failure and a perpetual screwup? Then normal brain tells you to put on Emmylou Harris and eat a lot of chocolate chip cookie dough ice cream, and play Boggle on the smartphone until your fingers fall off or until a fresh breeze, a stray compliment, or an inspiring quote from the ever-growing pile of self-help books beside the bed lifts you out of irons.

The trick is to figure out a way to bring disaster brain's unflappable logic—the philosophical equivalent of map and compass and common sense—to ordinary life. Because normal life doesn't have a clear view of the end of the road. Normal life is much more of a bushwhack. You don't get to see what's up ahead, or how to get there, and obstacles loom much larger than they really are or disguise themselves as worthwhile destinations of their own. So you let them lead you astray or you walk part of the way around them but don't make it all the way back to your original course, and the few degrees' difference leads to a different place entirely, one that makes you wonder, three miles, or a decade later, just where you went wrong—if wrong is in fact the direction you went.

I walked the Harriet Line two and three times a week over the next few months, as winter turned slowly to spring. That walk became a kind of companion to me—one that was never the same two days running, one that always had something new to say and remained consistently surprising, no matter what mood I brought to it or how bleak the weather.

By now, I used the compass only as a general indicator. I knew that if I traveled halfway up the rise of the hill and then walked more or less parallel to the crest, the broken spar of a tall pine in a small clearing would come into view, and from there I would see Jill's meadow and beyond it the smoke from the chimney of Harriet's house, and I would adjust my path accordingly. The return trip was even simpler. I would turn toward the cleft that the unassuming little stream that ran past my house had carved out of the hills a couple of geological epochs ago, knowing that sooner or later I would see the light flashing off one of the windowpanes, or catch sight of the blue of the solar panels off to the left.

There were other landmarks—a large rusted red box that I needed to keep downhill from wherever I was on the ridge; an intricate loop de loop of bright blue plastic tubing marking the place to turn right; an abandoned hunter's blind high up in an old oak that meant I was too far to the south. There were other signs, more ephemeral: a rust red stain in the snow, shaped like a peony in full bloom; two sets of tracks converging, but only one leading away.

Some days Harriet was home, and we visited, but much of the time she was gone. Dean had died just a few months earlier; the community church in South Woodstock had been filled with children and grandchildren and the many friends made by a man who had crafted a good long life along lines clean and strong. Harriet was a Yankee of the old school when it came to sorrow; she carried on. But her face bore a look I remembered, an expression no longer frozen in shock but drained of curiosity, the face of a soldier on a long march, intent on putting one foot in front of the other.

One day in late February, as I ducked under a strand of Chip Kendall's sugaring web, a small movement caught my eye. Perfect round circles were progressing very slowly through the blue plastic tubing. The sap was rising; sugaring had begun. What I was looking at were air bubbles caught in the sap, but their slow, pulsing movement, the drip of the transparent liquid, reminded me of the steady drip of the toxic liquids in the plastic bags during those long afternoons in the chemo salon, reminded me that I had survived to see this, one image at once superseding and silhouetted by the other. This was a tapping of life, yes, but life that would renew itself the way the spent body renews itself after all the insults it sustains in the name of saving it.

We all have our different ways of making it through the world, whether we consult map and compass or GPS or the trade winds. In Australia, aboriginal tribes travel thousands of miles through the country's interior desert by means of the songlines, as they are called in Bruce Chatwin's book of that name. According to aboriginal legend, the songlines are the paths created in the wake of the world's beginnings; they mark the ways taken by the creator during the Dreamtime. The people who followed those ways recorded them in songs and stories, dances and even paintings, and if you know the stories, then the songlines are the only maps you need. They describe the location of waypoints and destinations, and the landmarks along the way: lakes and rivers and the craters in the earth that are the footprints of the gods.

Some of the songlines are very short; others extend for hundreds of miles. Amazingly, speakers of many tribal languages can understand them because what matters is the rhythm of the songs, the melodic contour, which mirrors the terrain of the country traveled through—the words are almost irrelevant. To sing the song is to see the country it describes, to feel the ground beneath your feet.

I think we all have songlines, routes that are almost incantatory in their power to settle and soothe, to reconcile past and present. I still walked the Harriet Line long after it had ceased to be a challenge, and it nearly always brought me great peace when I did. It was a strange place that had become a familiar place, and, as such, a reminder, not always heeded but always there for the taking, of what can happen when you face a fear, take a chance, look forward and not behind. I learned a thing or two there.

Epilogue

In April, I returned to Castle Dismal after a month's absence in New York. The loft, which had been on the market ever since the market crashed, had finally sold, promising an end to the debt in which I had been drowning, but also raising a number of questions as well, all of which I had ignored as long as I possibly could. Finally, however, the closing date began to draw near—there were twenty-seven years of living to be taken apart and put into boxes.

Selling the apartment meant making decisions. Zoë would graduate in a few months' time. There was nothing tying me to New York now, save the usual rounds of medical tests and follow-up. But there was nothing tying me to Castle Dismal either, was there?

I thought about that. Something had changed in my relationship to my odd, ungainly house. I had begun to see things I had spent a good deal of energy not looking at. I could admit that I was sick of being alone, sick of the isolation and the difficulty. The strange thing was, the more I could admit to myself how much I hated life at Castle Dismal, the fonder I became of it. We were in this thing together in our awkward, cantankerous way.

Besides, life in Vermont was more doable now: I had friends, a routine, and more than anything, I had the plainspoken beauty of the place—not the overly quaint village and its green moneyed hills, but the round-shouldered grumpy ridges that surrounded but didn't shelter Castle Dismal, the scruffy dark interior splendor of the woods that covered them. They had given me a sense of where I was, a place on a map. Perhaps I worried that I wouldn't find that sense anywhere else.

And yet something was pulling me away. For a long time, I had been thinking about this question of where to live and who to be. I had learned, much to my chagrin, that I was not the self-sufficient anchorite I had thought myself. The solitude in which I lived for the last four years had been nourishing; had given me the room in which to unfold my cramped and crumpled thoughts and fears, to give them space to walk about and sort themselves out. But now, I needed people. I needed the world, a different world than what I had made for myself in Woodstock. A small town in Maine? Near my family in Virginia? The Isle of Mull? (A long shot admittedly, but the site of one of my favorite movies.) I tried to think adventurously, to imagine a place that could combine this solitude I loved with a community into which I could knit myself, to finally get it right. But my heart sank at the idea of once again picking up and moving somewhere new.

There was one place that did have a claim on me, though it took a long time for me to realize it. As I cleared out the apartment on lower Broadway, ran errands uptown and down, got in touch with old friends, the city, in its pushy insistent way, demanded my attention. I had fled Manhattan feeling like an exile and a failure, but now the city reasserted itself, clam-

orous, incessant, polyglot, struggling, a place where I could
turn a corner in a dozen different neighborhoods and find a
chapter of the life that had been my life, a place that was both
present and past. Home?

I wasn't sure. But I found an apartment, a small one, in a
neighborhood far from the one in which I had lived. It was near
Central Park—I was hoping that its open spaces would make
the move easier on Henry and, for that matter, on me. I would
rent it for a year, a place to roost while sorting things out.

A place for Zoë as well: she was graduating from college that
spring, and would live with me for a time while she sorted out
her next step. In a month, I would take my compass and measure out the distance from the place where I watched her walk
away that first day of her freshman year to the podium from
which the president of the college would hand a diploma to an
accomplished, confident young woman I was proud and grateful to know. One hundred and sixty paces, on a northwesterly
bearing of 320 degrees. A small distance. An immense journey.

I had only a short time to pack up the loft, and no time to
decide what to do with all of the things that wouldn't fit in the
new apartment, which was about a fourth of the size of the old
one. I took the coward's way out and had most of it carted off
to storage until Zoë and I had time to go through it all when
she was back for the summer.

The last night at the old apartment I sat in the living room,
stripped now of nearly everything that had made it home. All
our belongings were packed in boxes, anonymous and hidden.
I had been afraid of this night, of the storm of memory and
grief it would arouse, but in the end it wasn't like that. The
room was quiet, calm, merely a space now, one where I no

longer belonged. Yes, there was something irrevocable about leaving, but there was something light, something fresh as well. Lee was gone in a way he had not been as long as we lived in the place where he had lived and died. But so was a feeling I had never wanted to acknowledge, of being trapped in time, of being unable to escape a long obsolete idea of who I was.

When a man or a woman dies along the Bajo Urubamba River in Brazil, the family does not stay in the house where he or she lived. Sometimes, they burn the house down. But to live on the river is to walk a thin line of survival, so more often, the place is taken apart when the family moves away, and the thatching is used for a new roof, the doors and window frames placed in the walls of a new house in a new place.

There is a powerful reason: the *samenchi*, as the Piro people call them, the dead souls, refuse to leave their old homes, so attached are they to the life they lived there, to its joys and sorrows, to the people they loved. If you stay in such a place, the Piro believe, the *samenchi* will haunt you, weeping and begging you to join them. The living must turn their backs to the seductive call of the dead, no matter how hard this may be, or they will die themselves.

After the last box had been packed and the movers had come, I had a moment of alarm that Lee would not be able to find us, that he would think we had abandoned him. It was an embarrassingly illogical thought, but nonetheless it lingered for the first weeks in the new place as I edged my way around mountains of boxes, trying in that first wave of unpacking to find sheets and towels, spatula, teapot, mop and broom.

But then I opened the first of the many cartons of books, the ones that had surrounded us for so many years, old dusty hardbacks, many of them collected by his parents, a few first editions he had saved for and collected when he was young. One by one I picked them up and brushed them off and found a place for them on the bookshelf, and that was where I found him, where he had always been, as he had always been, a part of the conversation between the present and the past, a cherished part of life. But not, in this new place, the only thing I could see in the room. I had wondered, I probably always would wonder, if I had let too much of my life end with Lee's death. Maybe I had. There were still times when I missed the company of men fiercely, and the daily intimacy of married life, not to mention sex. But when I thought about love, I thought about this: my husband was the only man who had ever seen me for who I was and didn't blink, the only one I had loved the same way, the one who survived the myth I first made of him, and let me in. We had been able to be ourselves with each other: that was all. It takes time to love and to be loved like that. Time and luck, I suppose. It seemed a little greedy to expect it to happen again.

I thought about the list I had made, that first fall at Castle Dismal. Deal with Sex, I had written. Figure Out How to Be Old. Had I done these things?

When I first moved to Woodstock, I had been afraid of wanting something that, as I aged, would be in increasingly short supply. After treatment had ended and my hair had grown back, I tried dating once again, in part to prove to myself that I still could. Some of the men I went out with were in-

teresting, but nothing really took. Dating was more an obligation than anything else; I didn't need the reassurance it used to
provide, that I was attractive or smart or interesting, and what
I did miss about relationships were all the things that came
only with time—shared glances, and private jokes, the kind
of physical intimacy that glows long after the heat of the first
flames dies away. Dating, with its elaborate explanations and
rote retellings of old stories, was tedious and, especially on the
Internet, a lot of work. I didn't have the pluck or the optimism
or the capacity to deal with rejection that my bolder, more successful friends did, and my profiles all seemed to convey that
message, according to friends who vetted them.

It occurred to me, as I sat across from the kind, nervous
veterans of bad marriages and disappointing children, that I
might never have sex again. That was too bad—there were
days when that was really too bad—but I realized it wasn't the
end of the world. Trollope observed in one of his novels that
the reason some men grow avaricious as they age is that avarice is a passion compatible with old age, something to focus
on when other passions fall away. I couldn't see avarice suc-
ceeding eros, but I did find pursuing desire at this stage of life a
little awkward and unbecoming.

Besides, time and solitude had taught me that it wasn't love
I wanted nearly so much as a life, a full one, made of friends
old and new, and a generous curiosity about the larger world,
where the keen edge of desire could be deployed in less narrow pursuits. Into such a life love might accidentally stroll, but
whether it did or not, it could be a satisfying life and a useful
one. And for me that was more easily done in the mutable and
marvelous city.

As to the other items on the list, the only thing I could say for certain was that my ability to decode Ovid was improving—slowly. Growing older, now, that was trickier—at this point, I was still grateful to have a shot at getting old. Cancer teaches you to appreciate the ordinary: the bright yellow teapot on the gray winter morning, the patter of rain on the windowpane. These are the things that save you.

I had wanted aging to be an adventure done with panache, even an art form. That was hooey, I learned. Getting old was not a business to be romanticized, and as for accomplishing it with grace, that was as easily done as sailing through adolescence without a gawky moment. But I had come to look on myself with more equanimity and, occasionally, even kindness. Regret was something I had less time for—the mistakes I made no longer loomed large as true north at the top of my map to the past, but had begun to diminish—most of the time—into the cliffs and swamps, the dead ends and roundabouts that over the years had altered my course one way or another.

The notion that getting older meant a chance to discover who I *really* was had been alluring. It seemed to posit that all the mistakes I had made as a young woman were merely a passing madness of the blood, not faults of character cut into the living rock. But that idea had its dark side as well: if everything you had done or wanted was a product of biological and social destiny, of the urgings of the womb, then what was left when the storm subsided, when the flood tide ebbed, taking with it the thing on which you had pinned most of your identity?

As we age, we hear ourselves repeating our parents' refrain:

if only I had known then what I know now. But perhaps the reverse was true as well—perhaps what we miss about being young is all we *didn't* know, the mystery and terror of being young and clueless, the capacity of life to surprise when there was so much of it left to discover. I thought I had needed to know what was next, to have a game plan, but now I wanted only to go forward not knowing exactly where I was heading, but ready to look about me, as I did in the woods, and to see what had been there all along.

The last time I walked the Harriet Line as a full-time resident of Vermont, it was April, the end of mud season, and small lavender wildflowers were beginning to show themselves through the muck. The snow was gone, so the landscape in which I walked was very different now; I was ducking under the sap lines I had once stepped over.

Henry and I reached Harriet's house more quickly than usual—Henry in particular was anxious to see her because he had found a thoroughly disgusting deer skull he wanted her to have. She came to the door with an apology in her eyes: she had visitors from Massachusetts. Could we come back later?

The day was fine, so we continued on, up to the end of Noah Wood and then into the woods. We followed the stream for a while, splashing through its twists and turns, and then took off in what the compass said was the general direction of my house. I rarely took a bridle trail or footpath now.

I wandered for the sake of wandering. The spring came late to these hills, and as yet they were all still scruffy, a monochromatic brown, bare boughs only hinting at coming glory in tightly furled buds, the young pines barely visible among the trunks of fallen giants. Still there would be something, a shaft of light, a precocious bloom that would take me one way and not another, and so I kept a watchful eye on the thin red needle that pointed toward home.

We emerged from a thicket of brambles and there before us, a little sooner than I had expected it, and farther west, was the old pile of mossy stones that marked the beginning of the grassy road that would take us the last part of the way. It was a road that used to drive me crazy; it was the most direct way back to Castle Dismal, and yet I could never find it on any map—I would stand on this road that was clearly a road and look for it on my topo map and it wasn't there. It says something about the uncertainty in which I lived then that I believed the map over my own two feet and followed the road with a great deal of suspicion, half convinced that if I didn't watch it carefully, it would take me somewhere else entirely.

Maps, I know now, are not static. Walk in a place long enough and you see all the mistakes that have yet to be corrected, the disconnect between the three-dimensional reality on which you walk and its two-dimensional representation. Walk in a place long enough and even the most accurate maps fail to represent what is actually there.

When I look at a topo map of that patch of woods that I so painstakingly learned to navigate, there are no obvious errors, but neither is it a map of the Harriet Line. It's just a uniform square filled with irregular shapes and sepia squiggles, indicating a ridge here and a road there, a small creek, and a handful of steeps and valleys. But to me this same patch of ground is a dense and intricate landscape, a place both mysterious and comfortingly familiar. I know it inside and out, and I don'-

know it at all. It is a place as big as California, as slender as a birch, a place that has borne witness and given rise to mistakes and understanding, a place where I have barked my shins, and crowed with delight, and wandered dazzled down a highway cast by a fat old moon. An ordinary patch of wilderness, to which I am most beholden.

Just as the woods are more intricate than any map could indicate, so, too, was the past more complicated and less drenched in emotion than my crude line drawings—this is where it all went south, this is where my luck turned—would indicate. Wander around in your memories long enough and you begin to realize that the maps you've made to the person you were and the life you lived can become outdated; my personal map was based on a set of narrow and harsh coordinates: all that I didn't do or failed to accomplish, everything I meant to be and wasn't, all the good things that had been and were no longer. The result was a map that failed to account for most of the country it covered.

"Any life when viewed from the inside is simply a series of defeats," George Orwell observed. Most lives don't add up to much on paper: a child raised, a task or two well done, too many people disappointed, one man well loved— accomplishments that may not mean a great deal in the eyes of the world, but rich nonetheless in both beauty and squalor, in the mortality of the red bloom on the snow, the transcendence of the bright bird on the wing. Quite a big deal, in fact, if ne who did the bushwhacking that brought you to e you find yourself.

rote in *Walden* that he went to the woods in order ely, to pare life down to its essential facts. I went to the woods to run away, to begin again, to become a strange and fabulous creature: my true self.

I'm no longer certain such a thing as a true self exists: we are all of us a web of genes and circumstance, of accident and purpose, and our notions of identity are as entwined with time as they are with blood and bone, nerve and sinew. Besides, the question no longer mattered the way it once had: getting older is largely a matter of getting over yourself, of stepping out of your own way, the better to see the world through a wider lens than the narrow preoccupations of self had ever provided.

I wasn't any of the things I had strived to be, or tried to escape. I was just a walker in the woods, who had learned a thing or two perhaps about finding her way, one who would get lost again and again. With luck, I would walk into the future the way I walked into the woods, with my wits about me, with curiosity and humility, with a first aid kit and a compass.

Or so it seems to me now. We believe what we need to believe, in order to get on, until life takes its next swing and we land, on top of the world, or brushing the dust from our knees, and once again, we make ourselves new maps.

Acknowledgments

This book owes everything to my extraordinary editor, Jennifer Barth, of HarperCollins, who was the unerring guide, navigator, and sharp-eyed observer through the wilderness of my first drafts, always able to see not only the forest and the trees but also the way going forward.

My agent, Philippa Brophy, was, as she has always been, a lighthouse on many a stormy night and a constant friend and source of encouragement.

Thank you to the people of Woodstock, Vermont: to Susan Morgan of the Yankee Bookstore; to George, Linda, and Josh at the Village Butchers; to the staff of the Woodstock Pharmacy; and to the amazing women of the Whippletree Yarn Shop—Andrea and Shelley, you kept me sane.

Thank you to Kathy and Jon Peters of Runamuck, especially for Henry.

For the information on the Piro Indians, I am indebted to Peter Gow's essay "Land, People, and Paper in Western Amazonia" in *The Anthropology of Landscape: Perspectives on Place and Space*, edited by Eric Hirsch and Michael O'Hanlon.

To Garrett Epps and Joe Olshan, there can never be enough hosannas for the comfort, kindness, and confidence.

Harriet Goodwin and Lynne Bertram are my guardian angels and good friends—thank you for taking such good care of me.

Thank you, Cynthia Gorney and Megan Rosenfeld, for your reading and your advice, for your good humor, and for being two of the most dazzling women I know.

To Peter and Susan Osnos, as always, deepest gratitude for your friendship and your kindness.

To Doctors Ronald Ruden, Bonnie Reichman, and Alexander Swistel, and to the nurses and technicians of New York–Presbyterian Hospital and Lenox Hill Radiology—I would literally not be here without you.

And finally and always and in every way, thank you to the radiant and redoubtable Zoë Lescaze, my love, my happiness, my dearest child.

About the Author

LYNN DARLING is the author of *Necessary Sins*. Her work has appeared in the *Washington Post*, *Esquire*, *Harper's Bazaar*, and *Elle*, among others. She lives in New York City.